GW00692152

DESIGN *& Intuition*
Structures, Interiors & The Mind

C. LEWIS KAUSEL

C. Lewis Kausel

Interior Architecture, School of Design,
Mount Ida College, Massachusetts, USA

Published by

WIT Press
Ashurst Lodge, Ashurst, Southampton, SO40 7AA, UK
Tel: 44 (0) 238 029 3223; Fax: 44 (0) 238 029 2853
E-Mail: witpress@witpress.com
http://www.witpress.com

For USA, Canada and Mexico

WIT Press
25 Bridge Street, Billerica, MA 01821, USA
Tel: 978 667 5841; Fax: 978 667 7582
E-Mail: infousa@witpress.com
http://www.witpress.com

British Library Cataloguing-in-Publication Data
A Catalogue record for this book is available
from the British Library

ISBN: 978-1-84564-574-8

Library of Congress Catalog Card Number: 2011922773

No responsibility is assumed by the Publisher, the Editors and Authors for any injury and/or damage to persons or property as a matter of products liability, negligence or otherwise, or from any use or operation of any methods, products, instructions or ideas contained in the material herein. The Publisher does not necessarily endorse the ideas held, or views expressed by the Editors or Authors of the material contained in its publications.

© WIT Press 2012

Printed in Great Britain by Polestar Wheatons.

All rights reserved. No part of this publication may be reproduced, stored in a retrieval system, or transmitted in any form or by any means, electronic, mechanical, photocopying, recording, or otherwise, without the prior written permission of the Publisher.

Contents

Foreword

The subject taken up in this book harks back to the days when Cecilia Lewis was a young undergraduate student, a topic which has continued to fascinate her to this day. With her broad knowledge of history of architecture, her talented hand of an artist, her sharp eyes of observation and her keen imagination, over the years she has discerned myriad connections between the built form and the human form —not only in architecture but also in clothing and headgear. In this monograph, Cecilia expounds and elaborates on some of the reasons underlying these parallels, which in her view are not just coincidental and the result of mere structural resistance — i.e. form follows function. Instead, she discovers profound emotional and philosophical reasons for why some forms seem recurrent if not eternal, in particular those which relate to the overall body shape, and especially to the head. In this book, she makes her case not only with the aid of numerous photographs of actual constructions, objects and paintings, but also with original pieces of art in her own hand, with which she demonstrates graphically some key ideas underlying her interesting vision of designed forms.

Come and explore with her the fascinating world of design and discover the Leitungsmotiv in architectural form, i.e. the motivation psychology behind the choices of form in homo faber.

Eduardo Kausel
Professor of Civil and Environmental Engineering
Massachusetts Institute of Technology

Preface

People delight in both art and artistic processes, and are compelled to travel the world to visit sites and collections that belong in the heritage of all humankind. In Design & Intuition: Structures, Interiors and the Mind, Professor C. Lewis Kausel analyses the public enjoyment of design. The author constructs a comprehensive vision of design through an approach that focuses on the visual as well as the sensory impact of forms and materials on human creativity and experience. The book looks at the continuity of ancient structures from development to the migration and transformation to ornamental design. It traces the development of repetitive motifs observed in historic architectural structures, interiors and artifacts, and analyses the contemporary inspiration in the look and design of the commonly used mechanical instrument.

The author brings, from obscurity to awareness, a group of interiors which resemble anatomical forms which seem to be a fairly unintentional outcome of creativity. Buildings and humans at times reflect each other in structure and form. The design phenomena featured in the book is to a large extent, unnoticed in the basic everyday experience of design. As the author reveals there are apparent veiled relationships which are present in certain important architectural and crafted works around the world. Important among these are the many designs that show architecture as a surrounding form for the human figure or head in medieval sculpture and vaults, and they approach a complementary body form as seen in the trefoil shape. While architecture can be evoked in the design of people's appearance in the clothes they wear, for social ends and other subconscious cultural dynamics, human anatomy can be echoed in architecture as well, which is probably an intuition of mirroring the self that surfaced in design in periods when its practice was not self-conscious. This is a mind phenomenon that seems through thought and the subconscious to connect people and buildings in little-known perceptual relationships. The book reflects a cultural dynamic in the making of design that can be studied by the variables that affect people collectively, in the way that people sit, move and be. These provide a basis for the explanation of the phenomena observed in this book.

The observations, detailed illustrations, content and arguments of the author add new valuable cultural dimensions to the basic knowledge of design. From here, an interesting possibility arises, where a cultural dynamic of the human mirroring its self sometimes transforms structures making them approximate forms that evoke outlines of the human body or the cloaked human form. This is an interesting logic of a visual type in these forms. The presence of the human outline in architectural structure may be a simple reflection of cognitive habituation to it, as the person moves through space and the architecture evolves to cloak it. As well, the design of the surface of the brain may reflect itself in architectural covering structures as seen in the medieval vaulted ceiling. These repetitive elements of structural support have been woven into a supportive web of stone tissue. This anatomical form is present in important structures of history having materialized, it seems, intuitively.

These processes are different from the pure Classical proportioning of architecture after body measurements. Classical proportioning is an ideal form far from the natural state of the phenomena observed in this book. Common vocabulary falls short of the true dimension of the author's observations, which is a fascinating discovery of the mind's intelligent but unknown functions. The exploration of the book's intuitive designs compels a vision of the cultural mind as seen through design. These observations also point to a phenomenon observed in philosophy, namely that in developing knowledge the mind sometimes arrives at solutions that parallel the solutions of the natural world. In history architectural elements were often merged with other configurations of organic form creating a fluid visual transformation from structural support to inventive life form. Humans concretely reflect themselves in designed form.

In the cultural activity of designing, symbols and shapes are used and transformed. Historic costume mimicked architectural forms, which transformed plain appearances, and enhanced the complexity and sophistication of societies. The representation of structures can also be found in the crafted object such as miniature arches, which the author sees as depicting not just ornament but parts of renowned buildings, whose forms are a figurative language that human culture perceives. Her book offers fresh new interpretations of this fixation of culture with certain architectural form. A key research tool here is the fact that important products of human creativity have been more lasting than the life span of their creators. Society holds fast to favorite architectural works, and though the contemporary world is busy enjoying innovative and highly sophisticated, often simplified design, the attraction of ancient, sometimes complex, aesthetics is actively sought today. Upon contact with the contemporary psyche images from the past elicit a response to a familiar form, as architecture has been and is an instrument of expression to which the human responds. The study of design on the collective psyche has been necessary for a long time. The input of the public is vital, and an important missing piece in the picture of what creativity is.

Based on the above observations this study was developed gradually by C. Lewis Kausel, starting over three decades ago. A key achievement of this book is to point out the importance of culture on the making of design, to extract and explain an important previously-unknown relationship of the human brain to the interior of architectural vaults, the outline of the upper body and trefoil arches, and a deeper look at certain designs whose generative concept seems to be living form themselves. Design and Intuition puts together the domain of the visual and otherwise sensory dialogue of the intellect with that of design: historical intellect built and solved successfully an intuitive logic based on imagination and perceptual experience; that of the body's dexterities, the natural world, the experience of forces, equilibrium, the behavior of materials, fantasy and environmental harmonies.

Design and Intuition: Structures, Interiors and the Mind, evolves further the author's themes of previous publications, and adds many other examples of intriguing architectural forms with corresponding versions in crafts and in functional industrial design. The book develops clear visual connections between architecture, the human form, objects and costume, which the author sees as based in the collective visual experience of world culture. This book is a doorway to a promising way to visualize an important domain of creativity: the transforming possibilities of design to culture. It will interest designers, researchers and insightful lay readers who appreciate unforgettable design. The book's findings will likely interest researchers working on studies of the exchange of mind and matter. The author's research and methods of study are firmly grounded in objective observation and cultural phenomena.

Kathleen Driscoll
Associate Professor, History of Modern & Contemporary Art
Director, The Gallery at Mount Ida College

Part I:
INTRODUCTORY ESSENTIALS

Introduction

You may have observed that some designs outlive their creators' life and are in style well after they were conceived. Why should this be so? While both, historical and contemporary designs enjoy public favor; the so-called traditional or classic forms have a long record of application, and hence, permit to observe some of the cognitive constituents of design that can lead us to the study of this and other associated characteristics of the relationships of design and the mind's intuition. In fact, many well-known design images have been around for a very long time, although the processes that have kept them in use always offered various types of versions of them and adaptations. Design advanced in history, and surely newer concepts and ideas displaced older ones, but inspiration and analogy continued many old and ancient forms that never went completely out of style. Some design patterns have survived for millennia with visually recognizable characteristics, despite the fact that recycled design images are tailored to newly-arising standards.

The reapplication of structural and ornamental forms has been a characteristic habit of various cultures. Certainly the development of an effective method to build produced in history a design configuration that was distinctive and was detectable in the buildings that resorted to it, and these features came to influence the style of the crafts. New methods and instruments are created in culture that are applied and multiplied. It has been easier to repeat a system that works well, rather than design a different one, and this axiom, except for exceptions, is still applicable today. Reapplication of parts and systems is promoted by the building trades of carpentry and also by the finishing practices, and not least by the understanding of design or ideas that move forward the profession. However, practices and persuasions do not in themselves bring about the long term attachment of society for certain images. So to these practices we must add the inclinations of clients for certain design. In fact, the design that keeps people's interest is what primarily makes architectural types thrive and creates niches for the practice of certain trade skills. Then, the observable permanence of some very old forms seems to be driven as well by the attraction they can exert on connoisseurs, aficionados and amateurs.

Behind the lasting or 'iconic' design patterns that people seek and cultivate, there is a kind of 'linguistic' quality in all visual culture, which we know in part as tradition. Sometimes certain known forms are perpetuated because they facilitate a communal 'fit in', something that probably works similarly to the song that everyone knows and can sing, which is popular, not because it is good music, but rather because it permits the merriment of a shared social activity. However, the advantages of easy and straightforward features ultimately operate against people's interest, hence, though exposure enhances acquaintance, the design configurations that have endured for centuries, must do so because of other, more important cognitive-aesthetic reasons that maintains people's interest.

The visual power of ancient images still reaches many people. Certainly most design should be interesting to viewers, and a certain amount of this interest is the suggestion of design. Design can suggest its age and its salient characteristics and issues. This is indeed a very vast subject and much has been written and said about the work of designers, especially architects. It is a topic as vast as the literature about the work of prominent artists and designers. This particular work focuses on the suggestive power of a few ancient designs that are still influential today, even though their influence thrives mostly among the smaller and residential projects rather than among prominent designers, hence, the popularity of these designs is public-driven. With this we don't want to promote historicist design, but instead study its phenomenon.

With regard to what ancient monuments were intended to suggest, we can start by briefly acknowledging just a single well-known fact: the design of history that was meant to be monumental, often manipulated building scale and adornment to offer an impression of stateliness and power. Ancient architecture was intended to inform the grandeur of a civilization in those who saw it. The goals of rulers were incorporated in it and gods were also propitiated and extolled in such monuments. It is interesting to point out that the intended social objectives of ancient buildings are however not immediately noticed by all who see them, especially for a first time, as for anyone new to the appreciation of architecture, some explanation or reading is required to grasp the social purposes of design. But contemporary people respond with interest to the visual character and forms of some ancient buildings, prior to learning about their social details. Thus, the ancient builders, for example, succeeded in impressing an unforgettable design in viewers and less so in the communication of their intended goals. What is exactly this substance or connotation of some ancient monuments that the historical mind expressed and the contemporary mind perceives? This is the central topic of this book.

Some basic structures have tended to become 'deformed' in the processes that reproduced them over a period of time; such as the elongation of a spire, or the pyramidal forms; or the sinuous forms as onion domes, or the spiraling elements and the complex rooflines of some traditional architecture. Their forms advanced beyond the configuration of basic structure, into 'further transformed' designs that we must considered to be cultural forms. A mutual feedback between the cultural psyche and the form of architecture seems to have existed behind the slowly-shaped distortions of structures. There must be visual rationales possibly in the culture in addition to those of builders, for some architectural forms to be distorted in such special vernacular ways. Evidently no one in particular guides a vernacular shape that takes centuries to evolve, but ultimately these design relationships are all aesthetic transformations (of imagination) so to speak, applied to the outcomes of methods and materials. The structures that were perceived as exceptional additionally had good chances to become décor in the crafts, as earlier mentioned here, and even to enhance some items of costume, moving away from buildings and thus reflecting that the public has enjoyed them and adopted them for applications external to architecture. The outlines of these well-known forms may become represented in some head attire too. In the mutual evocation of head attire and architectural tops, there is a well-developed cognitive visual phenomenon that connects architecture and the mind. The connotative substance of architecture would therefore be behind it, as well as a tendency of the mind to reuse the forms it finds fitting for some other uses. In fact, the observations done in this study indicate that design (and imagery in general) means more in culture than the fast-passing trends of fashion. It also means more than its socially-intended objectives. What needs to be realized is that there are cognitive circumstances that cause these adaptations of design, i.e., the phenomena by which the configuration of rooftops pass to head attire (both of which are above the head). Creativity is common to all human beings, and not only an aspect of an artistic profession. It involves a search for fitting forms and solutions in any individual who becomes interested in design. The mind of both designers and non-designers alike seems to have cognitive expectations in other words. We might suspect that these expectations are highly variable. However, in a few buildings, design has succeeded in communicating something quite fundamental, at a level that has not involved words, which the imagination of individuals, for instance, builders who lived centuries ago, shaped, and this substance became expressed[1]. Individuals of later centuries continue to respond to these buildings.

Internet today abundantly offers replicas of ancient art and this market recycles and sells 'historicist' artifacts[2]. This thriving market did not suddenly appear in our electronic age, as it has been in operation for a very long time, but Internet has made it more easily available. Furthermore, in the recent decades building products with classic styles have been mass-produced and are abundantly found in stores. This type of production is the embodiment of the success of images that are built beyond their original context.

So the reader is hereby invited to this book's search for a holistic understanding of design that takes into account the public perception and response to forms. We shall approach this topic through the migration and permanence of design in the crafts and the cognitive relationships that can be observed in this phenomenon.

---

[1] This book offers several examples of this type of design and are discussed and illustrated in its chapters.
[2] Those that look like history's designs.

Foreword to Part I:

The Endurance of Design Against the Odds

The permanence of design is not a recent event, and it is observed in the public demand for certain styles. Design images are always being cultivated by persons from all walks of life hence this 'keeping' of old images is not a simple trend. There is ample evidence from ancient times that Greek builders consciously reproduced the plans and details of archaic temples, and ancient stone buildings featured carved ancient woven patterns and wooden structures. This is a case of surface representation of ancient techniques. Images of structural pedigree are also discernible in the crafts that society has collected and treasured in museums, and in the industry that replicates them. The millennia that have elapsed, reflecting that people keep and repeat the most desirable features of architecture, for whatever reason, indicate a tendency in culture to do this, and this collective, perhaps 'custody' of the mind, that doesn't forget nice old forms, has been noticed in the social sciences and in design[3]. The images of structures, then, are associated to something that was significant when used as configurations in the crafts; but could they mean much if they are not necessarily recognized symbols? See for example, a reliquary with a Byzantine dome, pediment, columns and roof in Fig. 1, in which an architectural image has been reproduced in the object. This example illustrates the replication of a building's dome, arches and roofing outside the architectural context of these structures. The appearance of this type of building has migrated to the miniature. In the case of this particular Byzantine church, this event must have been helped by the aesthetic quality of Byzantine architecture, as the overall design of volumes and the decoration of these churches can make an object appealing as a well-crafted work of art. But additionally to Byzantine aesthetics, the representation of churches in reliquaries was a particular craft that was not bound to one specific architectural type or manner of execution. The image of a church seems to have been advantageous to a church object and/or needed in the congregation in some meaningful way. Since the equivalent forms in architecture and object, are not created in the same way, or with comparable means, our common design vocabulary (created for building elements) can be unsatisfactory to refer to the similarities observable between the two. But the example of this reliquary permits us to attempt

---

[3] This has been seen as conservatism, nostalgia and as a tendency of culture.

an initial description of some qualities that are perceived in both a building and a miniature. To start the topic of the transference of design as a motif, and its purposes, we see that the religious building indicates consecration (sacredness) to society, and if we focus only on this quality for now, the sole image of this architecture 'passes' this quality to the container.

One of the aims of this study is to understand the extent of the influence of building forms in the design of other crafts and the purposes and dynamics of this phenomenon. Structural support such as an arcade, or a column, can be an image reservoir that society reuses in other types of applications in design. A migration of the configuration of architectural elements is discernible in the design of interiors, artifacts and sometimes costume. This happens in all periods, including the modern, though in this latter one, the reflection of architecture in crafts or costume has been until recently –as today this effect is a fashion– less discernible than in history. The ornamental recycling of structures was an ancient practice reflecting a migration of the visual aspects of structures to other artifacts. The objectives of a structure are static in nature rather than purposely visual however, structures offered forms and elements prior to the freedom of form that we find in our all-is-possible era. Some structural forms have been reproduced for many centuries. It is indeed noteworthy that ancient design is still in use today, as this seems to reflect a long standing satisfaction –if not a visual fixation– of culture with some ancient images.

Architectural elements are first achieved by technical methods[4] and the visual power of the ancient static systems could seem promising for the cognitive acceptance of society of a world primarily made of technical processes, in which we don't bother to add embellishment; and such was a central design ideal of the twenties to the seventies. But these available outcomes should not lead us to conclude that a world of design free from the shaping of fantasy, or from human hands must automatically be good to the psyche, as society forms bonds to architecture, as described above, and the suggestion of design that the public in spontaneity, tends to support, will always have to be fulfilled. We know by now that even though there is variety of opinions in this sense, industrial processes by themselves may not always produce forms that satisfy the public's expectations. We will touch again this point when it is relevant to the discussion of our examples.

When the ancient arches and columns are applied outside their original context, as in giving classical character to a table, they are no longer a construction method but they are meaningful images to those who reuse them; that is, sometimes, a mere concept that is carved or engraved, or designed as a miniature superstructure, onto some other framework. The propagation of the meaningful structural images to the crafts is driven primarily by the fine execution of the forms observed in buildings, and their aesthetic appeal to society. People desire to possess or contemplate particular aesthetic examples. Hence, we see ancient colonnades, or distinctive doorways in a building, sized and framed by a favorite style; for instance the popular classical or Georgian pediments in English-speaking cultures, or in an elegant modern style that represents people's forward-looking values in fine design. These images may carry connotative values that become involved in the communication of visual ideas, such as distinction and elegance, for example[5].

---

[4] Though this refers to the tectonic architectural elements there are also ornaments in buildings which precede architecture that have been thought to derive from technical crafts (Semper).

[5] These images were resourceful to Post Modernism that tried to reutilize them, although the approach to design of this trend was not –or has not yet been– satisfactory to all.

Indeed, as we can see all around, architecture and many structures such as bridges; have been a suitable subject matter for a large amount of artwork by engravers, painters, photographers and the pictorial applied arts. Architectural elements that are widely enjoyed in society motivate the re-use of their forms, thusly feeding the recycling of a known, albeit modified structural image. While this cognitive mimetic event happens abundantly in culture, reflecting the visual substance of structures in other artistic applications of various kinds, the study of the reasons for these repetitive applications of images was not objectively tackled in the modern age, perhaps because the progressive viewpoints of the age discarded an inherent logic of imitation in a cultural phenomenon that echoes images of the past. However, whether inspiration and design emulation are imitation or not, is to some extent a matter of the vision about design that reigns in a period.

In the case of historicist architectural styles, this is not done anymore to apply a construction system that works well, or any other practicality. If a home is built today with period features and details, it is done because the public responds favorably to this design language. Buildings and interiors are destined for consumers, and hence, what has been maintained is ultimately what users look for and choose from the design available to them in the environment.

In a different type of craft, such as furniture, (see Figs. 2 to 14), the building elements applied to the pieces seem to impart distinction to them. Architectural elements may not only impart special aesthetics to crafts, but they can make a piece appear more 'intellectual' to some viewers, than other alternate décor[6]. The images of secular structures (including those originating long ago in a classical temple, but no longer associated to religion) contribute other connotations than the form of a church to interiors and crafts. In a classical temple, sacredness is forgotten to the mind's senses and is too far back in civilization; and all that columns and arches still impart is solemnity, distinction and the sense of firmness. We identify today the idea of the 'classic look' in the antiquities we still cultivate. Other features, as good materials, artistic level, craftsmanship and excellent finishing, are important too, and we shall come back to the more complex relationships that reliquary, architecture and furniture hold in later sections of this work.

The mind creates associative ideas that become widespread and customary in a period. Products for consumers often attempt to enhance appearance, or offer one kind of impression or another. Products and design are thus affected by collectively-perceived qualities. Distinction and elegance are aesthetic dimensions created in society that ultimately belong in certain sensitivities of the psyche. The reality of these attributes and qualities, which amount to the way the physical features of objects are interpreted, exists collectively in the visual transactions of people and relies on similar abilities in most people to perceive and understand them. As one person sees these features, many others are likely to see them too, or can learn to recognize them in a similar way. What elegance is, evolves with time, and varies in different regions, hence, what was elegant in the Rococo may be perceived as opposite to what is elegant in modernism; and because change in style does turn around the appearance of the world, the elegance of a historic piece must be measured, not by contemporary perceptions, but by trying to understand the aesthetics favored when the piece was built.

---

[6] To many people the styles chosen reflect who they are (or want to be). These are aspects of social and personal behavior which often become intimately associated to the world of design. Intellectual appearance can be hinted in subtle ways and often blends with the concept of elegance. Indeed our species is proud of its intelligence and so it seems to indicate sometimes in items of visual culture.

Furniture is an interesting functional design in that it is both utilitarian and artistic. Its scale is limited in most pieces by the height of the body and the reach of the hands[7], and the form of chair seats and backs, are designed considering the body's contact with their surfaces, or should be, as when they do so, they are best. Thus good furniture design is a function of these and other practical variables[8], but important pieces have always been executed with artistry, in whatever style they are created, and many artistic pieces have not been necessarily comfortable. Some historical pieces were greatly affected by the visual expectations of users, at the cost of comfort sometimes, as some thrones. Furniture is also expected to harmonize with interiors, and hence, the evolution of first class examples has been harmonious with the style of interiors, if not in the same style. Additionally, furniture is subject to repetition of pieces as it must provide serviceable sets of matching parts and identical seating. Yet, despite all these practicalities and expectations, furniture is a type of item that can elicit a collector's interest and can be a work of art; and since every household has furniture, a large segment of society is interested in it. What contemporary users want minimally in personal furniture is usually comfort and durability, and also craftsmanship[9] when affordable. Furniture is therefore a kind of instrument that we use, in that it can accommodate and extend our body and it helps the organization of our possessions and tasks. It is also best when it helps to raise the level of distinction of our reception rooms and is additionally visually satisfactory to our psyche.

The desirability of architectural styles in furniture over the course of history is a cognitive interest that speaks of the character or visual qualities of buildings that can be present in some furniture types. In this type of décor structural elements are represented in items of human scale, and this is important –though at a subliminal or unspoken level– for the here-mentioned substance of some architecture which is that hard-to-forget aspect of it. Though the forms of architecture are by no means the sole –or even predominant– décor that furniture manufacturers produce, the phenomenon of style, for instance, is strongly influenced by forms in high varieties of expressions that emanate from prototype architecture to other crafts, even in non-literal or subtle ways. By directly displaying structural elements, many furniture pieces may become weighty. However, this is not to define here that decoration with columns and arches is a synonym of superior work, as this involves surface touch rather than form-giving, a very important aspect of design. But above all, this judgment depends too on design execution, and certainly on the preferences of the beholder. But to this study, the perceived value of a piece by the public is certainly important, for its collective character and the value of furniture may be higher to the average consumer if a piece displays some special shaping than if the piece offers no suggestion to imagination. Many traditional pieces display the ever-present cornice-like tops with molding; others may have pilasters. Certainly a piece could suggest an animal form too. If anyone asks the furniture craftsperson today, why these features are reproduced in furniture, the answer is 'buyers like pieces with these features'. This type of ornamental enhancement by means of non-functional structural images is also common in traditional architectural exteriors and interiors[10].

---

[7] These limitations are derived from customs but are not determined by rules, as high shelving can be reached by a small ladder.

[8] Such as the dimensions that fit through doors, the objective of storing and the assembly of parts.

[9] This may seem obvious to the reader, however it may not be what the manufacture of institutional, educational and outdoor furniture focuses on, as this furniture has problems of maintenance, exposure to public treatment and fire safety issues that affect materials and aesthetics.

[10] While not all readers can agree with the indiscriminate use of pilasters and moldings to enhance the appeal of artifacts and architecture, the current home industry carries out innumerable renovation projects where pilasters and classical molding are added to previously plain buildings and are resold in a higher price range.

Design configurations seem to easily pass from architecture to interiors and furniture. This direction has been clearly dominant in the course of history, judging from the designs that have survived. It is visible in classic design. The mind's tendency to reuse design forms is involved, as well as the ways artifacts relate stylistically to buildings and other reasons of social nature. Instead, a form created for static purposes will normally not follow imagery, and before even considering any such possibility, it has to obey its own needs, which greatly determine the final outcome. However, in spite of this, it would be simplistic from our part to deduce from these facts that all migration of design without exception will always be unidirectional, i.e., from structures to objects, as we must always expect a few ingenious examples in which formal influences migrate from a humbler but beautiful craft to the higher one; or from nature to structure. We must also acknowledge that before structures were ever created, some materials offered their resourcefulness and they come in certain forms and natural applications that may 'give' ideas on how to use them to the creative observer. Imagination is unbound, and ingenuity can sometimes manage to shift a typical direction of influences, not to mention as well that a final product of technical origin can also allow the suggestion of other appearances. Examples abound in the shaping of technology for consumers. With regard to building examples that have allowed influence from other crafts, we can cite for example the mutual feedback of the appearance of fine porcelain and building décor. This is observable in some rococo plasterwork as that of Italy, France, Southern Germany, and in Adam style (see some examples in Figs. 119 and 120).

In a somewhat comparable way, to deduce that the images of functional design are sufficient for all design is also too one-dimensional, as previously mentioned. The mind has created fine decorative design, good enough to be treasured in museums, and these examples make such principle weak. However, we must acknowledge under this topic, the looks of some early industrial apparatuses such as an outdated telephone, the earliest sewing machines, and radios, which have become 'decorative' today when they have lost their functionality, and the public seems to easily grasp the interest of these forms. Such apparatuses are destined to consumers, thus no wonder they were somehow designed with a minimum of 'visual charm' in mind, and this makes these machines still interesting after they have become obsolete. Hence, what is exactly decoration? This is a crucial question, as there can be no borderline between embellishment and other approaches to giving shape. Some points about decoration will be illustrated in the examples of this book; hopefully shedding some more light to this paradoxical topic, and clarify a few of the weaknesses of an anti-decorative direction. Mankind has indeed spent great efforts in developing methods to build by using a statics-to-form design, but there are many examples of roofing, ceilings, domes and gables that were used as décor in fine furnishings. Furthermore, some structures continued the evolution of their form beyond statics and this transformed some of them into ornamental shapes, as the mentioned spires, Asian roofs, Eastern domes, for example. Why has this been done if not for the type of aesthetics that culture pursues? The 'distortions' of structures are vernacular syntaxes. To professional rationales that don't include the work of culture they have seemed a bastardization of 'true' design; however, these cultural developments obey intuitive criteria of society that in all likelihood are not purposeless, and even though we can educate the lay culture about our principles of good design –and we all know that laypersons want to learn from trained designers– in the overall effect, and in the long run, our principles may not change the spontaneous perceptual modes of people. The developments that happen in design without being purposely-intended can indicate the

We believe that this and other popular cultural responses to design are a very interesting visual phenomenon that deserves attention for what relationships they can elucidate about the perception of design.

presence of a collective or cultural perceptual domain. The levels at which design is grasped in this way by most people need attention and study. Our rationales also would be more universal if they were more inclusive of the perceptions of the public. It will always be tougher to try to change the intuitive visual aspects of the psyche than to study them. We need to understand first why the mind tends spontaneously to perpetuate design and why it utilizes it as it does. Then, we also need to observe the distortion that design undergoes when it advances based on pre-existing images. Are these distortions showing a direction, or are they haphazard? If, on the other hand, our formulations consistently leave out the spontaneous human behavior toward design, they can never show a picture of the value of design to people, hence to the world.

Is there importance in a design logic that transfers images from one artifact to another in culture? This event is surely tied to a dynamic of collective visual perception, and can receive only preliminary answers in an early approach to the topic. But, we can understand first, that this is the way visual culture works, and in this territory, we meet a universal world that should have some importance. Second, visual unity in the crafts (for example achieved by the similarities that all crafts in one style display) can also be cited here. The migration of design is involved in the stylistic character of an age or a world region. The identification of ages and regions in the visual character of design is a cognitive affair, and this must have its own measures and significance. But under any circumstances, the migration of structures to the crafts is a collective phenomenon and hence it must have a *raison d'etre* beneath its surface. If there can be an objective understanding of good design it is not only the study of what's in vogue, or of just innovative forms, which as much as the world needs them, they are only half of the complex picture of the phenomena of design.

The widespread cultivation of old styles in the world shows some basic levels of visual concurrence. There are likely other instances in visual culture of aesthetic agreement among many. Though most of us reject the idea of things in common, there are many situations that can affect how we all respond to some developments as in the acceptance or resistance of society to a change of circumstances in other aspects of living. This should not be too different in the world of design, except that it would affect primarily visual domains. For example, the public has not rejected a fast turnover of design forms in electronic apparatuses yet, and has been attentive to more innovative compact and seamless shapes in the looks of these items; showing that people enjoy the design improvement of new items. Indeed the world seems to love novelty and improvement of design, in particular after a design look has been multiplied too much and has become too common. But the same public is less convivial to changes involving items that have an established reputation, and also has reasons for keeping stable the décor that goes into important ceremonies. In fact, there have been instances in the history of design when a good number of people responded to design with a near obsession for it, and there have been others of cold reaction toward some styles.

Innovation is an important modus operandi of the field of architecture and interior design, and it became at least in principle, a design ideal of our times. Design education does not encourage designers to thrive by reproducing past styles, as it had been the case in the previous century, and the reasons are clear: first doing so produces stasis of design, and second, discovery, technology, and design are intimately connected to the advance of an age, and innovation occupies a central place in the practice of design[11]. But innovation also faces the production of much similar design on the other hand, as innovation is not exclusively practiced. In the reality of materials and methods, a majority of buildings and other types of design are based on existing methods and solutions. If

---

[11] Being important for the public too.

variation is possible, nobody likes housing blocks of identical buildings[12] for example, but still, at the opposite extreme, new design vocabularies are not a true norm, not even in progressive design, as new design is not created for each project. The human habitats are organized by both cognitive (visual) and functional logics, where doable and favored design circulates in the architectural practice and the crafts. Today, we explore the sustainable features of vernacular buildings, and this new direction makes our field focus again on pre-industrial design examples.

There are surely common grounds in visual perception, to which all people respond similarly, or we wouldn't see the same colors or be able to understand aesthetic dimensions such as gracefulness or elegance, but there are important differences too, that arise in personal sensitivity and degree of exposure to visual culture and to refined crafts. In creating a new style, the ways users will respond is an unknown today.

What is considered prevailing in design at a moment in time, has been a function of the phenomenon of fashion, that changes according to variables and ideas that arise associated to discovery, or in more random ways, by popularity or aesthetic pronouncements within the field that manage to influence society. However, some responses of people indicate that overall, many individuals are more intuitive than obedient to fashion in their choices, in the fact that the cultural sponsorship of favorite design often denotes proclivities for past design in people. The twentieth century schools of architecture counted on the possibility that the public can eventually become receptive to *avant garde* design, just as designers embrace progressive directions themselves. Hopefully this principle works, however, it is still not possible to know if this hopeful vision must always work, unless there is objective study of what already worked in previous outcomes in place.

Since design is often expected to fit personality, or convey a desired social or professional character, the amateur who knows sufficiently about quality in products, and how to harmonize them stylistically, feels free to follow his/her personal inclinations. Our age offers abundant images and options, thus, people select design that appeals in personal ways; hence status images, professional looks, or exquisite, romantic, fantastic and even anonymous design, are some of the many varied criteria that people exercise. If users invest in some designed entity, it should not go out of style after just a short period of time as fashion does, at least, not its intrinsic value. Opinions and explanations vary enormously; each person knows clearly some reasons for preferring some design, but this freedom of choice in the public has been clearly inclined for the designs of history in particular in the case of homes, even during the modern age, and among users who live in cities where design *avant gardism* is a matter of pride.

Finally, the weight of technology needs a few words. It must be mentioned that in studying old architectural and interior design, it becomes clear that in history, technical principles were always in operation, even in the most decorative styles, i.e., they were vital to the final forms achieved. At any time, available mechanisms must be in place too for any design that is meant to be multiplied for many users. This was so even in pre-industrial society when the quality products of history were most often presented in handcrafted ornamental ways. Despite this central role of technical principles in design, however, it is interesting to observe that all forms, utilitarian or decorative, just the same become converted into a cognitive pursuit by society. These relationships are not contradictory, or should not be so. They only need to be known and understood. Once people are well exposed to a certain product and can choose from multiple options, they very naturally take

---

[12] Though this housing is built in this way in economic housing.

the nicer product. We all certainly search for many products for their efficacy and cost for instance, but the looks and sensory qualities of design are important too; and when there are options of comparable quality and cost, the choice is surely the nicest. Sooner or later supply and demand inclines the balance in the direction of designs that become well-liked. This is also how customers come to establish common visual preferences in the market, and certain designs become popular. The long term success of any style becomes a function of user acquirement, and in due course, the maintenance and perpetuation of favorite products, is the final victory of design. And even if some of us may put down this dynamic behind the success of this design as 'crowd pleasing' and "popular taste' and think it has little worth (it is also a passive effect over design, as the public is not involved in designing), at least, we should not fail to notice that the public interest and attachment for certain products is what drives markets, and this is immensely consequential to production. It is what keeps feeding the perpetuation of traditional architecture and interiors and antique replicas as well, and this interest needs no help from organized advertisement for a new look, to keep living on. It only needs that an industry of crafts offers these designs. These cultural dynamics are observable even if few things seem more subjective than the human response to images. Antique forms are ancient concepts that have become more permanent than centuries, ages or empires, and there must be cognitive reasons for this. While to some public this is tied to habituation, to many others it is compelled by honest aesthetic preference. The result is that ancient forms still live in the twenty-first century and it must be because of something they suggest to the collective psyche that maintains them in circulation. The average public responses to design images lie unattended, and in dire need of understanding. But despite of this requisite, history's favorite styles must be balanced with innovation. The antique manufacture by itself, without alternate options, would surely cause design to make no further progress hence, the effort to create innovative looks is always necessary in the overall phenomenon.

The enhanced understanding of theories of design promoted more loudly in education during the recent century than in previous history the aesthetics of technique and function in design[13] and this, of course, has included industrial outcomes. In the early days of a budding industry there was experimentation and not all outcomes seemed alluring to the public, as they seemed too different from previous forms to public perception. In the everyday practical application of design, if the objectives of fashion have not been easy to digest to the larger public, both designers and clients found agreement on some solution after some communication of objectives. Social behavior acts as 'glue' as we well know, and makes a tradition-oriented layperson and a designer –who would rather work within innovative lines– find common ground. Social variables also operate in designers themselves hence their wish to create something that a large audience can enjoy should be active all the time. Social necessity leads to cooperation, and people of different inclinations and opinions tend to compromise and concede. If a designer's output touches cognitively both the public and his/her colleagues, it wins the approval of both laypersons and adjudicators, and this is ultimately a road to professional success. However, despite the strong influence of social pressure (and also advertisement) the quality and merit of design are not a function of social behavior. By this we mean that though social behavior 'acts' on design, talent is not dependent on it. Finally, social behavior can also act sometimes against creative expression[14]. Thus the thrust of this study is not exactly the characteristics of design as a function of social behavior, but more precisely, the suggestion of design forms on the mind.

---

[13] There are exceptions to this however, as in some deconstructivist architects, where not even technique should prevent the freedom of an architect to shape a project in extraneous ways.
[14] Through ascetic views, inhibitions related to luxury and even aesthetics and the like.

Architectural Elements in Artifacts and Furniture: Design that Transcended a Structural Context

(a)

Figures 1a Illustration by author, 1982 and b (overleaf): The umbrella dome as an architectural motif of a reliquary.

(b)

Some structures, such as the dome, gable roofs, arches, columns and pediments have been found suitable to be reproduced as artwork, as the highly sculptural example shown here. The construction of elaborate miniatures focuses on the appearance of a special building. Behind this motivation is the human desire to treasure or contemplate a particular architectural image. In both artistic and vernacular crafts, transformed structural elements convey architectural character to other objects. The church form suggests the sacred character of the object.

This precious architectural object reproduces the "umbrella" dome such as that of St. Emmeram church in Regensburg. While this type of Byzantine dome derived from a ribbed system of gored segments first built for structural objectives, this dome type advanced in the direction of a decorative form. Under each arch there is a human figure. The scale and proportions of the parts are certainly ritualized. The container rests on metal eagles (not shown here).

The *Welfenschatz* Cupola reliquary of Cologne was one of two reliquaries made with this umbrella dome. The materials are bronze gilt on an oak foundation with *champleve* enamel and ivory walrus carvings. Its height is 45.5 cm and dates from ca 1175–1180. It was constructed to house the head of Saint Gregory of Nazianz. It is kept today in the *Kunstgewerbe* Museum of Berlin. The other similar reliquary is in the Victoria and Albert Museum, London.

Figure 2: Nineteenth century miniature Romanesque arches and columns in stair balustrade. *Museum of Science, London. Architect: A. Waterhouse.*

Figure 3: The furniture leg as a fluted column. (*Image from What Antique Furniture used with permission*).

The fluted column is a stylistic device of the French neoclassical style.
   Contemporary reproduction of a graceful and elegant French table by Adam Weisweiler, who made furniture for Louis XVI.

Figure 4: Wall cabinet with trefoil arches.

Lebanese, traditional. Cabinet is in current use. Top clocks tell the time in various parts of the globe.

Figure 5: Structural analogies in interior millwork. Pilasters, cornice and dado panels.

London pub. Interior is in current use.

Figure 6: Design for architectural Cabinet with building elements of the
Italian Renaissance. (*Image source: Schwenke, 1881*).

The architectural elements suggest the heaviness of an ornate neo-Renaissance style. Cabinet
shows fluted pilasters, dado panels and cornices at both top and under the squat colonettes.
The rectangular spaces for tryglyphs display lion masks. The top cornice displays dentils as
in classical buildings. The pilasters project to the front all the way to their lowest point, as in
late Renaissance architecture. The panels depict symbols representing the arts of sculpture,
architecture and painting.

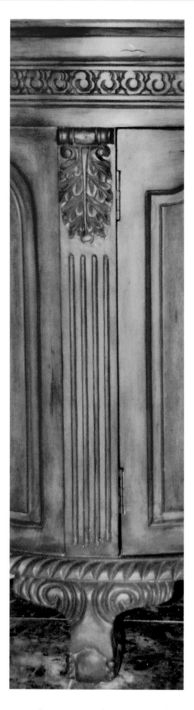

Figure 7: Architectural interior pilasters still featured in common
contemporary furniture and fixtures.

Detail of a sink cabinet in current use, 21st century. American.

Figure 8: Chair back with exquisitely carved architectural niche, Late Renaissance, 1580.

This *sgabello* chair is just a decade away from the beginning of the Baroque in Italy. The niche evokes an Italian Renaissance wall. The chair back is inclined and gently curved. Design by Giovanni Maria Nosseni, born in Lugano (today Switzerland). Stained pear wood, carved, gilded, painted, inlaid and polished. Stones are jasper and serpentinites from Erzgebirge, a mining area. Serpentinite was used in the royal residences of Saxony. Its color varies from greenish gray, to reddish brown and grays. August I of Saxony, summoned the Italian architect Nosseni to Dresden and asked him to create 12 chairs and a table, using serpentinite surfacing. The backs of four of these chairs have bust portraits of Roman emperors, including Julius Caesar, and Otto VIII (Holy Roman Emperor). Three others are simpler, of a later date. The ornate legs have semiprecious stones. The factory of the Walter family produced them. Only seven chairs remain from the set in the *Kunstgewerbe Museum* of Dresden. Rendering by author from a chair featured in Ernsting.

Figure 9: Cooke and Wheatstone double needle telegraph shaped as classical building. *Artifact in the London Museum of Science. 1843DSC_3002.*

(a)

(b)

Figures 10 a and b: Church *prie Dieu* and lectern in current use; with Gothic tracery and arches.Ely Cathedral's Lady Chapel.

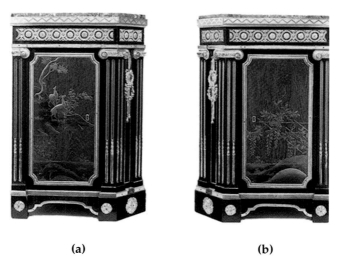

(a)   (b)

Figures 11 a and b: Cabinet with pilasters and Ionic capitals at corners.
(Image from What Antique Furniture. Used with permission).

Reproduction of Parisian antique curio stamped by Joseph Baumhauer. 1765.

Figure 12: Florentine architectural *Secretaire* cabinet. Sixteenth C.

Open cabinet displays profuse building forms that evoke decorative late Renaissance interiors or Mannerism. The structures portrayed prominently are the arch and pedimented facade.

The Florentine Renaissance and Baroque workshops used walnut, ivory, ebony, *pietra dura*, alabaster and bronze. This *Secretaire* is in the collection of *Chateau D'Usse*, France.

Left: Reproduction of an English throne chair. This baluster was used in English Baroque, Early Georgian and Colonial interiors and entrances. Image from *What Antique Furniture*.

Right: 19th C, detail of a drawing for a cabinet, Berlin *(Schwenke)*.

Figure 13: Furniture balusters designed after the columns of the baldachin of St. Peter's, by Bernini. (*Used with permission*).

Figure 14 a and b: Arches in wood ornament doorways in traditional interiors. London, England.

Chapter 2

The Life of Culture in Design

Everybody probably can relate to the impact of attractive and well-kept urban areas, the impressive structures, or the delight of being in a magnificent interior, and even the strong desire to possess an object of interest. We need to see this 'ability' of some design to elicit our strong approval, because what is experienced with interest is remembered.

The field shifts its dominant paradigms as new visions and design possibilities offer novelty and some advantages. New design parameters may displace older features and circumstances. Users who cannot maintain an old condition may miss an old design and feel nostalgic. This is basically to miss something known (familiar) that is gone[15]. Nostalgia (homesickness according to some dictionaries, or recollection and memory according to others) is frequently thought to account for people's proclivity for traditional ways and styles[16]. A nostalgic feeling can be elicited by coming in contact with a place previously enjoyed, or by moving out of a location after years of living there. In addition to nostalgia, the appeal of old looks is also tinted with romanticism for the lifestyle and comfort of another time. People develop fondness for places and attach high value to possessions. Change can end some comforts and impinge on customs once exercised. Comparably, if the face of neighborhoods is altered, it can affect users and bring up nostalgia for the way it was before. It can be hard for anyone to see their childhood house being demolished and replaced by an altogether different building. Human emotions can thus be affected by urban renewal and the replacement of buildings, even if it is just the effect of associative events. Similarly, it is sad when a precious old object loses its original integrity, or is stolen or lost. The general public wants, collectively, to keep structures that often were not even meant by its builders to last, like the attractive Alhambra of Granada that its builders constructed in fragile adobe and brick, and its preservation is difficult. People identify historic design as a cherished commodity, and many protect old architecture as they protect endangered regions and species that are threatened by extinction.

---

[15] The many features in an environment that are identified as 'familiar' are sometimes said to be stress-free.
[16] An early and thorough explanation of nostalgia is offered by the historian of religion Mircea Eliade, 1969.

For the sake of simplicity, the longing for styles past, and all associated moody states might be understood as a form of nostalgia, which is a feeling that clearly enters in the attachment and design picture; however, it is not sufficient to identify nostalgia as a culprit, nor say that it causes the wish to perpetuate design. More knowledge is necessary to comprehend the connections of society to old design. Some buildings are esteemed by many people, and perhaps the ties of all generations to the best built examples reveal a connotation that the psyche detects in architecture, that must be ultimately expressed in it. It is hence, important to understand how or why some design elicits this response that resembles fidelity in people. Many historicist images have never been personally experienced by the Western public who admire them, such as ancient Egyptian, Chinese and Aztec architecture, and vice versa for the non-western public who have not experienced the Loire chateaux or the cathedral of Florence –just to offer a couple of examples out of hundreds. There are multiple issues involved in this actual cultivation of architecture, and nostalgia does not account for some of its key cognitive phenomena. The world's monuments elicit cognitive connections in all kinds of people, and not just by hearing, reading or learning, but their magnificence and syntax is also grasped through photographs and film. Thus, even though nostalgia *is* a cognitive mover, to this study other ideas have proven more workable than nostalgia, such as the *authority*[17] of ancient images, and connotative content of design and its suggestive power.

For instance, the migration of structural forms to artifacts and costume as featured in Figs. 1 to 37, is not clearly related to nostalgia. It can be observed that there is an analogous synchrony of configuration between some structures and head attire that could derive from the connotations that society (i.e., the cultural psyche) perceives in structures, and tends to represent in other articles, such as costume and crafts (see Figs. 15 to 38). Individuals have clothed themselves with the connotative value of certain structures, just as someone who puts on a wedding dress or army clothes dresses as bride or soldier. The idea behind it must be to try to connote something that buildings can connote (though primarily subliminally-achieved and certainly not done for pantomime.) Thus people perceive a kind of iconography in structures that may be adapted for personal looks, somewhat as the adaptation of structural motifs in the crafts to raise their importance or significance. Hence, some ladies of the English Renaissance surrounded their faces with the symbolic outlines of arched gates and gables, and even roof tiles. French ladies in the fifteenth century adorned their appearance with the iconography of the spire, and by this they connoted goals associated to the significance of a spire, possibly associated to a vision of the church[18]. The entire movement of a slender body with a very tall cone causes a different motion of the body that perhaps was perceived as graceful (see Fig. 21) and the importance of a church in the fifteenth century urban landscape that stood taller than the rest of buildings is favorable for a lady's appearance too, not to mention the value of religious virtue in the age. These fashions have been made possible by exposure to these forms in the immediate environment. People have resorted to special aesthetic aspects of these structures and wanted to evoke them in their appearance.

Turbans enlarge the top part of the head and hence probably can subliminally evoke the mind's powers as those of religious sages. These mental powers have always been highly desirable and

---

[17] This concept has also been discredited by those who have been critical of the Post Modernist arguments. However, there is little vocabulary available that hasn't been subject to negative judgment.
[18] For example, there is a reference in the Scriptures of the Church as "the Bride", and the symbolic implications of this understanding have been covered in iconography. See for example Schiller.

create respect[19]. For instance the mythological Chinese god of longevity Shou Hsing is depicted with a very tall forehead and large head, reflecting how the mind of worshippers adjusts the normal physical features of the head to indicate some particular characteristic of the god (we presume here, his superior gifts.) Religious leaders of some faiths may have a ceremonial head cover that parallels the top of their religious buildings (see Figs. 19 and 20), and this kind of aesthetic parallelism is a cultural component or feedback that affects architecture. This feedback in turn possibly 'distorted' the domes of some world regions, making them progress in the direction of some head attires –or should we more accurately say that both domes and attires were deformed to evoke a tall forehead? Turbans and onion domes then have a cognitive connection to each other. This connection may be the visual indication or appeal of the mind for its high abilities. Costume is derived from completely different resources and ideas than architecture or furniture, but the outline and configuration of some structures can be associated to the human outline through costume design, and this promotes the parallelism of their configurations. Certainly the character of a building can impart solemnity and ceremonial quality to clothes as it does to the crafts. In turn, the human appearance may have sometimes influenced some of these structures, but certainly not in their static conception, but rather in their subsequent cultural adaptation. People have suggested some of the elements or outlines of architectural forms in themselves, perhaps to appear more holy, or elegant; or dignified, as in the case of a ruler or a ceremony leader; or to stand out in height (or in importance) as a spire stood in the medieval urban landscape. The enhancement of human appearance through the recollection of structural forms, or the enhancement of domes through cultural shapes, is observable in many geographical regions and periods hence this is a phenomenon of the cultural mind and design.

Inspiration in old architectural forms is happening all the time in any artistic or design medium. A full blown revival also happens. While many examples of design migration make use of the influence of architecture in one and the same region and period, others resort to the visual authority of a well-established much older design. Traditional images are present in culture and are ubiquitously cultivated in art and antiques. Hence, these images are always part of the built environment. Every new generation of artists and designers will want to give a try to creative works inspired by ancient architecture. Good examples are found in the market of antique reproductions. See Figs 2 to 14, which revived the configurations of previous structures and styles. This means that an offspring design may appear much later in time than the parent form. This takes place because the connotative life of design becomes independent from its originating age by being present in all periods. This effect that can span centuries is an important dynamic for a researcher to notice, so as to understand the cognitive reach and freedom that is characteristic of design. Form analogy could be produced by a craftsman by seeing history's designs directly in historic buildings or by seeing and studying them in drawings and paintings, as many of the Palladian, Neoclassical and early nineteenth century Egyptian revivalists did[20].

In a somewhat baffling way, there are also some elements in architecture that carry in them surfaces that evoke the tissues and other features of the human anatomy, which are achieved (likely in

---

[19] Other examples involving the communication of respect and status (rather than higher mental powers) can be cited, as the tall wigs used in the ancient Greek theater to represent roles of authority, such as kings.

[20] For example, Raphael Sanzio was an early artist with archaeological interests who studied ancient Roman monuments for his high Renaissance design; Inigo Jones studied the work of Palladio and Scamozzi; Christopher Wren studied Jones, Palladio and Bernini; and Robert Adam and Thomas Jefferson studied ancient Roman monuments to create their Adam and Jeffersonian styles.

(a)   (b)

Figures 16 a, b and c: Shallow architectural cone and Thor's hat. The hat may emulate this ancient roof. Author's sketches.

Left: One out of many examples of conical roofs. Ruin of the medieval Teutonic Knights, 1207. Cesis, Latvian SSR.

Right: Teutonic god Thor with his hammer *Mjollnir*, bronze figurine, Reijavik Museum, Iceland, AD 1000.

This shallow conical roof form has an ancient origin as indicated by the Etruscan tomb below. It is observed sometimes in Early Christian churches of Italy, Greece and Armenia. It is common in Romanesque towers of Germany as in Saint Gereon, in Cologne.

Left: Conical roof in stone in ancient Etruscan tomb. Meyer's Lexicon, 1902.

Ref: author's unpublished paper, 1977.

(c)

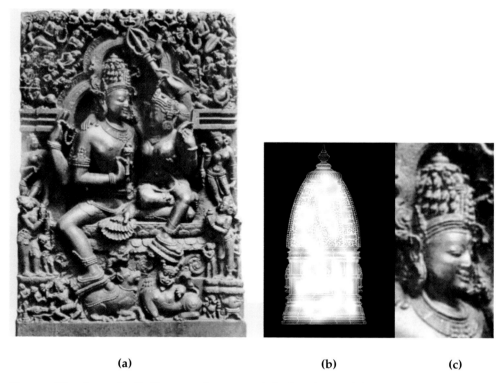

(a)   (b)   (c)

Figures 17, a, b, and c: Indian temple tops and the head top of deities evoke each other.

In Asia, sacred monuments such as stupas and temples are echoed in various forms in sculpture and the culture's crafts, such as ceremonial head attire, ritual dance attire and parasols. The varied work of different sculptors show some differences of height and details of the head attire, but the temple's outline is frequent. In literature about Indian religion, the features and artifacts that gods wear are carefully described, since they represent cosmic symbols and powers. Vishnu is described as wearing his own long hair gathered around his head. The mind arrives at formal analogies in cognitive ways that can be independent from materials or linguistic description. The temple form and Vishnu's head arrangement are cognitively related by their configuration.

Left: Sculpture of Shiva-Shakti, 10th century AD, in the British Museum.

Center: North India's Temple typology called 'latina'.

Right: Shiva detail.

(a)

(b)

In Asia, pagodas, stupas, temples and parasols are echoed in various forms in buildings and the culture's productions, such as dance attire.

Figure 18 a: Fifteenth century Temple of the Emerald Buddha rooftops, Bangkok. *Credit: internet photo mikaul.*

Figure 18 b: Thai temple of the Emerald Buddha sculptures with roof top head attire that echoes the roofs of the temple.

Figure 18 c (overleaf): Costumes of the sculptures of temple sentinels. These sentinels are depicted as supporting the cornice with their hands. The similarity of the sentinels to the roof tops could speak of an idea that the temple has an animated aspect. This is reinforced in the architectural like figure below, whose head cover is a structure like a cupola with a texture that evokes roof tiles.

Figure 18 d and e: **Architectural sentinel.** *Phrakaewfigure.* Temple of the Emerald Buddha. Sentinels are representations of the temple, but with a living form. The parallelism of the elements and the surfacing of temple and the attire of this sentinel is astonishing.

(c)

(d)   (e)

Figures 18 a,b,c,d and e: Parallelism between Thailand temple roof tops and ceremonial costume tops of the same ceremonial contexts. *Source for figures 18 b and c: Internet same site as above. Internet. ramakien3-cc-photo mykaul.*

Photo credits: Internet sentinel-cc photo -elbisreverri. **Ref :** Author's unpublished paper from 1977. This parallel between temple spires and head attire is mentioned in costume books as well.

The Bishop, Florentine school, c. 1300.   **Beauvais cathedral nave, 1225–1573.**

(a)   (b)

(c)

This is a pair of forms composed of two parts of complementary outline related by ceremony. The Gothic arch is echoed throughout a cathedral and in its sculptures. The miter was developed from the papal tiara and was in use in the 11th century. It had a squatter and triangular form however. The tall and curving proportion acquired by the miter may have been influenced by the development of the Gothic vault, as the development of several new religious orders and the design of vestments of religious orders is reported in the middle ages. The miter, although earlier than the Gothic vault, offers a possibility to become transformed in such a way as to approach the contour of the nave. That is, the miter's peaks were suitable to create a match with the nave's ribs and so it was shifted to resemble it.

(Continued)

Figures 19 a, b and c: Bishop and cathedral. The shape of a bishop's miter recalls the structure of the Gothic vault. The proportions and shaping of the miter helps the evocation.

At some point in time, the peaks of the miter, previously worn on the sides of the head, were shifted to the front and back position, as is still in use today. Also, the earlier triangular form changed into the taller and curving outline of a Gothic vault. The miter's resulting similarity to the vault is most likely unintended. A viewer (and the liturgical vestments designer) may have been only aware of a harmonious set of forms, and although he/she senses that the form carries substance, it remains as an unspoken connotation. The morphological representations probably symbolize the heavenly nave and its sheltering function above the high priest's image. The match of interlocking evocation tacitly identifies bishop and church. The exaggerated elongation in the two designs speaks of a greater mind. The Gothic vaults have also reflected the anatomy of the head in some examples of England.

Ref: Lewis Kausel 1980–82 graduate thesis. A drawing of this parallel was published in the Journal of IFRAA of 1983, and discussed in the *International Journal of Design and Nature*, 2007.

(a)   (b)

Figures 20 a and b: Russian bishop's head attire and early type of Russian cupola evoke each other's form and outline. *Both sketches by author.*

The Church of the Savior at Novgorod (right) shows a typical ecclesiastical Russian style. It developed in the 12th century. The tall head attire of the bishop, as the church, has a cross on top as well. Designs of these types lead to some of the unexplainable connections of the mind to design where the bishop's miter indicates the church and the church, the bishop.

Left: Nikon. Seventeenth century religious leader of the Orthodox Church.

Right: Church of the Savior. At Novgorod. *Author's sketches.*

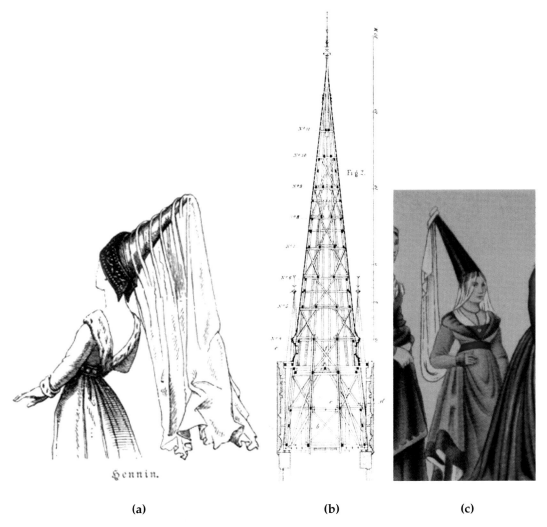

(a)   (b)   (c)

Figures 21 a, b, c, and d (overleaf): Church spires were evoked in fifteenth century ladies court dress.

The spire developed in the twelfth century to cap a tower's top. It became longer and pointed obeying a dynamic progression that was not necessary structurally. It moved toward greater height and slenderness and in this intent, the roofs that were transformed into spires transcended their original purpose. The spire's long and pointed form has been explained as symbolizing the heavenly aspirations of pious people. In medieval times a spire was a spectacular form that stood out in the urban landscape. It reached an impressive height in the cathedral of Coutance, in France. The spire is a cultural form rather than a structural one. Its genesis may be related to other significant cultural shapes. **Kennin**. Fifteenth-century head cover of France, Germany, Italy and the Netherlands. The contour of this head dress is proportionally similar to the shape of church spires. In evoking the form of a spire, this conical form also transcends the building context of a spire. This type of head cover makes the user taller and slender. The recollection of the church may evoke religious connotation desirable

(Continued)

for women in the period (perhaps virtue or devotion), however this is a symbolic form whose genesis is complicated. A tall pointed form was (and is) observed in the attire of penitents in Easter processions, in some regions. There is a painting by Francisco de Goya y Lucientes that shows a procession of flagellants with conical tall head covers. It can also be seen in Italian theater and comedy attire. Conical pointed hat forms are ancient however. They were used in Rome, made from flexible materials and there was a textile version of the form as well, that perhaps came from ancient Mesopotamia. This ancient pointed textile hat was similar to the Franciscan hood that also acquired an elongated pointed form in the Middle Ages. It was designed to mostly hang on the back from the user's neck and the pointed tip became elongated in the Middle Ages. This is observed in the paintings of Francisco de Zurbaran. See below. However, the medieval creation of the more rigid kennin as court attire offers parallelism with the spire.

Left: From Meyer's Lexicon, 1902. Center: Spire from Warth. Color example; Meyer's Lexicon, 1902.

Ref.: Lewis Kausel, 1982. Parallelism of conical form and spires is mentioned in *The Mechanic*.

(d)

Detail of a monk in a painting by Francisco de Zurbarán *Miracle of St. Hugo,* in the Refectory of Seville, 1655.
The pointed cowl is also observable in Zurbaran's *Saint Francis in His Tomb,* a painting of another miracle; that of the ascetic saint's apparition above his coffin, contemplating a skull in 1449.

(a)   (b)

Figures 22 a and b: Gentlemen tunics evoke architectural pilasters. 15th century tunics.

Early pilasters can be observed in Pompeian building doorways, but they were popularized to the widespread use we see in the Early Renaissance by Alberti and Brunelleschi.

Behind what seems to be a Renaissance fashion, there is additionally the idea of a pilaster as a representation of the human form. The human figure is subliminally associated to a vertical support (see Section 6 of this work.)

Left: Pilasters of Michelangelo's *Porta Pia* in Rome, 1565. Detail from a photo. Right: Pilaster-like tunics. Detail of a manuscript reproduced from Quintus Curtius, from 1480. Flemish, in the Bodleian Library, Oxford.

A crown with fortress towers appears in representations of the goddess Cybele (in Phrygia, Asia Minor known as Cybebe and Great Mother of the Gods), to whom the crown was ascribed by artists and poets. Cybele's crown characterized the cities of the world over which she presided. There is a bas-relief of the goddess in a Roman tomb. The ancient soldier who first scaled the walls of a besieged city received this crown as a reward (A. Rich, 1893.)

Figure 23: A crown is a symbol of supreme authority that evokes architecture in the *corona muralis* design. Some crowns have ritualized crenellated towers and turrets. *Corona Muralis. Meyer's Lexicon.*

Figure 24: A king is crowned with a crown, a head attire like a miniature fortress.
Source: Sketch from Lewis Kausel's graduate thesis. It reproduces a manuscript from the Cathedral of Lodi. The crown is held above the king's head and right in front of the image of a distant castle, more or less depicted above the king's head too. If we ignore perspective this fortress is too above the king's head. The contour of the crown evokes a fortress' towers. The positioning of crown and castle in this drawing is likely intended for the purposes of indicating the meaning of the crown. The peacock has various symbolic meanings, but one of them is royalty.

Sketch by author that reproduces a XV century Italian Manuscript.
A King's crown as a symbol of authority. The crown evokes the form of the fortress seen at a distance. The peacock is a symbol of royalty and has its own head adornment.

Figures 25 a and b: Hat and gown that evoke the crenellation of towers.

In a tower crenellation once facilitated the protection of warriors engaged in warfare with arrows. The crenellations around a rooftop became in time a visual device of religious iconography that indicated a city. In a hat and gown it has a similar function to urban identity. It evokes the general rooftop of prominent buildings as institutions. This parallelism could be achieved by a hat and gown designer who wants to convey a certain urban reminiscence for status and identity in the costume.

Left: Sketch drawn from costume book (1980.)

Right: crenellations in Palazzo Vecchio, Florence; Early Renaissance. **Ref:** *Lewis Kausel 1982.*

Figures 26 a, b and c: Italian Renaissance cap that brings to mind the mausoleum of Theodoric's shallow dome.

Painting by Bronzino. This evocation is one of general proportions and the exterior ribs of the mausoleum's dome.

Left: Mausoleum of Theodoric, 520 Ravenna.

Ref.: *Lewis Kausel unpublished paper, 1977.*

Tudor gable and dress of royalty. This gable head attire was also worn by Elizabeth of York as reflected in her tomb effigy[27]. It is reminiscent of some pointed gables; fifteenth century. The design of this architectural appearance seems to suggest seriousness and earnestness.

Left: Generic northern European steep gable.

Right: Head attire after a National Portrait Gallery painting of Margaret Beaufort. **Ref.:** Lewis Kausel 1982 and 1986. *Author's sketch 1982.*

Figures 27 a and b: Gables that frame the faces of ladies.

Figures 28 a and b: Arch, and lady dressed as Tudor Arch.

This Tudor arch coif[28] was used by Catherine of Aragon, Mary Tudor, Jane Seymour and many ladies of the court of Henry VIII, as seen in the drawings of Holbein at Windsor castle. The arch and this headdress are related in time and place. There are many levels at which patterns and details resemble between architecture and human adornment.

Socially, the perceptions of a period exert pressure on the way a woman must present herself. Architecture was a source of several head attire designs of late medieval ladies in England. This English Renaissance arch could have strengthened a Tudor identity (by association) in queens and daughters of Henry VIII (as their legitimacy was challenged.)

Left: Tudor arch, mid sixteenth century.

Right: Catherine of Aragon in Tudor Arch head attire. Sketch after a portrait of Catherine of Aragon at the National Portrait gallery. **Ref:** Lewis Kausel 1982 and 1986. *Author's sketches.*

---

[27] This head attire is referred to as Tudor Gable headdress in Britannica.
[28] This term is used in Britannica Encyclopedia.

Figures 29 a and b: The design of helmets and cupolas.

Helmets and domes are both a concave cover above the human head, in different scales. This 'silent' or unnamed protective meaning connects these distinct man-made forms to each other during most historical times. These forms acquired segments that evoke those delimited by ribs in some cupolas. They also have a characteristic ornamental top that could evoke a terminating lance point, or finial. A pointed tip in a helmet is however as ancient as the Iron Age. In ancient Rome the point became known as "apex". Its origin has been thought to be a pointed piece of olive wood that was set in a flock of wool or fastened to a cap [Festus s. v. Albogarelus. Serv. ad Virg, A. x. 270].

Left: cupola by Antonio Petrini, Wurztburg, Germany Right: Lithuanian helmet. **Ref.:** unpublished paper from author, 1977, and 1982. A relationship between ancient helmets and domes is suggested in E.B. Smith, 1950.

Figures 30 a and b: Roof tops and windows designed as helmeted sentinels.

There are vernacular buildings in Germany that evoke helmeted sentinels.

Left: author's photo of vernacular house. Environments near Weimar.

Right: Medieval Helmet of crusaders of Europe, Meyer's Lexicon.

This cupola by Asplund recalls the type of hat shown on right and so do cupolas of Parisian kiosks of about the same period. The lids of some classical vases and other serving objects may also show resemblance to a cupola and a beret. Certainly they are also covering devices belonging in an analogous functional logic of common experience.

Ref.: *author's unpublished paper, 1977.*

Figures 31 a, and b: Turn of the century architectural cupola and beret evoke each other.

Figure 32 a and b: Historic Saracen helmet and Iranean cupola show similar outlines.

The Shaping of Artifacts as Structures

Figure 33: Staff used today by the Archbishop of Canterbury, with fluted columns and arches.

Each arch has a religious figure in a niche.

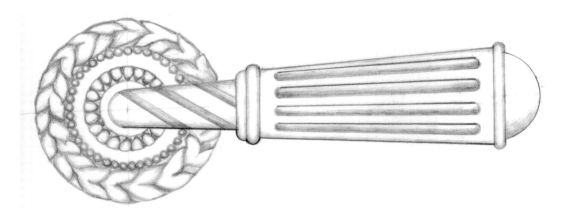

Figure 34: Drawing of a door handle with fluting that evokes a column.

French, Neoclassical, eighteenth C. This interpretation of the use of classical décor is perhaps by a hardware artisan who did not know how ancient Greece applied these motifs. *Author's sketch.*

Figure 35: Pictorial arcades as wallcover pattern. Mid eighteenth century.

The decorative arts have resorted to arches as an aesthetic image throughout human history, sometimes in pictorial murals and paintings and sometimes as a textile or wallcover pattern. The eighteenth century in England used textile wallcovers before the advent of wallpaper in the last quarter of the century. Damask and satin could have been used in the walcover featured in this painting so as to attain the silvery sheen of the colonades shown. Painting by Phillip Hussey, England.

Figure 36: French purse showing Gothic gate in its metal closing device. Fifteenth century. *Metropolitan Gallery.*

This is a fancy application of architectural elements in miniature. The architectural form and the animal head with a ring give everything of interest to this otherwise plain bag; French, fifteenth century. To contemporary eyes this miniaturized gate is romantic and evocative of an age of chivalry. Author's sketch.

Figure 37: Antique clock shaped as a classical building with six columns, pediment and dome; in use, London Pub.

Figure 38: Ring of King Charles I of England. Portrays a simple gabled building image and a skeleton. Meyer's Lexicon.

Eric Seale

Part II:

ANCIENT DESIGN: IMPLEMENTS AND IMPRESSIONS OF THE MIND

Foreword to Part II: Ingenuity, Intuition and Myth

Every human age before ours romanticized ancient times, and we may also fall for the looks of ancient periods. As we contemplate pictures of ancient stone buildings they seem interesting and we acquaint ourselves better with the visual language of such distant forms. The efforts people took to carve stone temples by hand and the ceremonial memorials constructed reveal a solemn mood that is rarely used in design today[29] but with which our mind somehow connects. The imagery of their architecture is curious, if not mysterious. Some ancient buildings offer that semblance or impression of something else that our conscious state may not see clearly, however, the vision of a building that sparks our interest remains in our memory and we feel inclined to photograph it, or to own books where a structure of interest is featured.

Earlier ages were no better than ours even though the ancient creative genius reached very far. There was no development of knowledge and people achieved their impressive buildings relying on their own natural gifts: primarily ingenuity, intuition and myth. They understood little of the forces of nature, or the origin of life, or the causes of diversity in species, yet these natural aspects of the mind built the basis of human civilization. In fact, intuition allowed Aristarchus of Samos in 270 B.C. to envision and to demonstrate that the earth circles around the sun. As society began to rely on scientific knowledge the exercise of intuition became inhibited by the rise of confronting attitudes and ideas demanding proof. However intuition remains the natural ability we have to identify correctly many (if not all) relationships or situations. Of the three natural aspects of the mind, mythology is an interpretation of the way nature works that reflects some internal understandings or 'memories of the human race' (archetypes), which are externalized in symbols. The ancient mind needed to rise above confusion and thus constructed systems of interpretation that explained the surrounding world[30]. Myth, though an inaccurate system, helped this goal. Early interpretations gave stability to society, necessary to begin to build the foundations of knowledge and eventually displace old visions. However, what matters significantly here, is that the natural abilities of the intellect, in raw or early conditions, became expressed in ancient architecture. Our ability to respond to them permits us to connect intimately with ancient design.

The perishable human state must have been continually evident in ancient times, and this may have nourished in part the solemnity, the search for permanency and the mood of pathos of classical design. This lasted up until medieval times. At the same time, the intent of rulers to keep control of their subjects has been associated to the use of awe-inspiring monuments and trappings, whose magnificence could make the masses diffident and respectful. Our contemporary world of worldly buildings is very different indeed, and so is our life in the midst of portraits of good times in our media and advertisement. One of the greatest contrasts between the ancient design and ours is that their temples elicit reflection on the spirit and the universe, as suggested in their design. Our age is not focused on such concerns, nor expresses them in urban monuments; notwithstanding contemporary individuals find them in the solemnity of ancient temples and mausolea.

---

[29] The sociology and anthropology of this mood in design is an interesting topic by itself, though it is beyond the scope of this study. It is a mood observable in architecture at different points in history, but not in all periods. It is not evident in the Rococo style, for example, but the application of the subsequent neoclassical style used solemn architecture again, in particular, in monumental government buildings. Today society has changed. Construction is business-dependent and has been motivated by selling, services and corporations, where other types of design have thrived.

[30] Our age still needs to do so and our dominant ideologies and beliefs are always systems of interpretation of the physical reality.

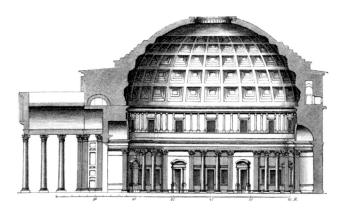

Chapter 3

Expertise to Impression

Buildings were constructed first in flexible plant materials, then in wood and brick, and in stone. The best examples offered stable prototypes to the view that showed a way to build; hence other buildings were constructed that emulated these models. Ancient prototypes that were built to last were primarily temples[31] and funerary monuments. The motivation of ancient agricultural societies to build large structures could have been related to the need to admit large numbers in halls at some point[32], which seems to have blended with a drive to display the largest structure possible as a sign of regional superiority. The oldest larger ancient structures extant today, are rectangular halls, achieved in stone, with massive pillars whose proximity to each other, of some nine feet of span, occupied much of the interior space. Ancient societies therefore sought to free the interior by a better method to span masonry over their halls, and they put into use the true arch. The clever arcuated system of spanning was an important ancient construction method. It can be found in Babylon, in the Ishtar gate (see Fig. 39), in sewers and hanging gardens. It was also used in ancient Chinese bridges and in the storehouses of Ramses in Egypt. This speaks of a very old origin and of a widespread application of arcuated systems that was both utilitarian (in the great strength of the semicircular arch), and aesthetic, as it is evident in the blue gate of Babylon. Centuries afterwards, arcuated forms captured the imagination of Roman builders with an incomparable construction development that expressed concave spaces. Rome built large groin vaults, apses and a few large domes. Before Rome, a

---

[31] The word *templum* is an Etruscan term that alluded to an area in the sky that a priest defined for interpreting omens. Later, this area was projected into the earth and a *templum* defined a ground consecrated to the gods. Often these areas had shrines. Gradually more permanent temples were built. The example from Etruria must reflect a development of temples elsewhere too, from the heavens to a natural site first, and later to a physical construction. The largest Etruscan temples that were later built were executed in wood with terracotta ornaments. Later, *tufa*, marble, cement, brick and travertine were used. Materials and methods must have varied in other regions.

[32] In some societies large halls first kept animals, while people resided in small nuclear units. This is the case in some Native American cultures. Some ancient temples, did not allow common people inside. However, halls are foremost the abode of human groups.

domical space is reported in the Odeon of Athens to which a poet paid tribute. For sure these statements that amount to celebrations of design, can be thought to have given direction to the general development of satisfaction and contentment of culture with these structures; but their service in spanning free interiors and their cognitive impression were surely very influential too, in their proliferation. The superior compressive strength of stone arches permitted Rome to put them in service in a wide spectrum of applications, and this was done to an extent not achieved before. Rome embarked in impressive architectural and civil projects (see the Pantheon in Figs. 40 to 43 and the slender aqueduct in Fig. 43). Concave structures seized the Roman imagination for long and had continuity in later styles. The resolution to the issue of spanning triumphed in the larger domes of Rome which were the temple of Minerva Medica (in ruins), the Domus Aurea (the Golden House of Nero renovated into other buildings) and the Pantheon. The Pantheon is a huge and advanced spherical monument of its age. The dome of the Pantheon was built with a 43.3 meter diameter on the interior, which seems to allude to its free space built as large as it was possible. It was faced with brick and the large dome[33], was made of concrete of varying aggregate composition, being heavier at the spring of the dome, the weak part of a hemispherical structure, and lighter at the top. The aggregate of the dome was graduated to attain a lighter weight as it rises and curves. It varied from a mixture of gravel at the base, to tufa in the middle and pumice at the top. The oculus at the center of the dome has a diameter close to 28 feet that seems to challenge its strength; hence, it is a virtuoso aspect of the structure, since in spite of being weakened by this window at the top it has not collapsed in nearly two millennia[34]. This could show a high level of control of the structure and its proportions by its builders, in particular because it has survived with integrity in spite of its age and Renaissance renovations (though survival in ancient structures has been sometimes thought as more 'luck' than anything else).

---

[33] It was ordered by Hadrian in about 118 – 119 AD, built to replace an earlier Pantheon began under Agrippa that is believed to have been a rectangular classical temple with colonnade, triangular pediment and gabled roof, from which the current portico could have been assembled.

[34] It is not possible to be sure however, since there was trial and error at this early stage of large dome construction. The builders surely sensed that the dome is supported laterally rather than at the oculus area — shown in the stepped thicker rings at the base of the hemisphere.

Although the arch's lateral abutment is provided by the flanking walls themselves, rather than columns and beams, the decorative representation alludes to the rectangle of forces of column and beam.

The structure is 12 meters high and is made of burnt brick. It is decorated with tiers of enameled dragons and bulls. It led to a processional avenue of more than half a mile long that was decorated with 120 lions *passant* and 575 dragons and bulls. The monument and its avenue led to Akitu House, a small temple visited by Marduk at the new year festival.

Ishtar was the great goddess of the Sumerio-Akkadian pantheon of Mesopotamia, identified with the West Semitic Astarte.

Figure 39: An aesthetic presentation of the utilitarian arch is evident in this ancient structure. Ishtar Gate, in the State Museum of Berlin, c. 575 B.C.

Figures 40 and 41: The Roman Pantheon.

Its present rotunda form was ordered by Hadrian in about 118 – 119 AD, built to replace an earlier pantheon began under Agrippa in 27 B.C. that is believed to have been a rectangular

(Continued)

classical temple with colonnade, triangular pediment and gabled roof, from which the current portico could have been assembled. The dome of the Pantheon was built with a 43,3 meter diameter on the interior, which seems to allude to a celebration of free space, built as large as it was possible to its builders. The Pantheon was faced with brick and concrete. The large dome, was made of concrete of varying aggregate composition, being heavier at the spring of the dome, the weak part of a semispherical structure, and lighter at the top. The aggregate of the dome was graduated to attain a lighter weight as it rises and curves. It varied from a mixture of gravel at the base, to *tufa* in the middle and pumice at the top. The oculus in the center of the dome has a nearly 28 foot diameter.

The oculus in the center of the dome is a design that was first tested in the Golden House of Nero.

The coffering diminishes in size as the coffered rows rise. The mid story with window-like recesses is part of renovations that were done from the Renaissance to the eighteenth century.

Painting of the 18th C. by Panini, goes back to the design showing marble surfacing, as it was prior to the Renaissance renovations.

Figure 42: The Pantheon's interior. Span is about 43, 3 meters.

Figure 43: The Aqueduct of Segovia, Spain.

The Segovia aqueduct was constructed sometime between the second half of the 1st century and the early 2nd century AD. It carries water from nearby mountains 10.6 miles from the city of Segovia. It runs another 9.3 miles before arriving to the city. The water is first gathered in a tank known as *El Caserón* and is then led through a channel to a second tower known as the *Casa de Aguas* where it is naturally decanted. The water then travels half a mile and the structure makes a turn. It is after this turn that the monument displays its full splendor. Its height of 93.5 feet includes 6 meters of foundation. The first section of the aqueduct was destroyed by Moors in 1072 and restored in the 15th century with pointed arches, rebuilt with great care so as not to change the original work. The lower-level arches have an approximate width of 14.8 ft. The aqueduct is built of un-mortared brick-like granite blocks. In the 20th century, the aqueduct suffered the wear of automobile pollution. Natural erosion of the granite itself has also affected the structure through the years. In 1997 restoration projects have been ongoing in order to guarantee the aqueduct's survival. During the restoration, traffic has been re-routed, and the paved route under the aqueduct is now a pedestrian zone.

Chapter 4

The Pantheon's Intuitive Iconography

History has recycled the trademarks of classical architecture in civic buildings and continues to do so in the details and artwork of primarily residential, hospitality and private contractor projects. Classical designs can evoke other forms and in order to study the allure of classical design today, we must be ready to explore what might lie behind its widespread world appeal by trying to understand how our inner psyche connects with the architectural design of antiquity. We should do this through the study of design itself rather than through the interests of social psychology, and the reason is that this is an appeal of forms to the psyche, that we must understand by the significance and aesthetics of design, more than as a function of the push and pull of social pressure. The next paragraphs prepare the reader to see the design concept that the Pantheon can present, even though this vision is not immediately evident.

Early societies sometimes depicted the representation of natural phenomena and common wisdoms by resorting to human forms. This was often an artistic way to express ideas, and many depictions show the *genius* or personified guiding spirit of a city which in time became identified as Athena and Dea Roma in Rome, and some other 'odes in marble' to desirable traits and to inspiration such as the virtues and the muses. Geographical features involving change, motion and energy that mankind's intuition must have sensed as 'energized', such as rivers, winds, the sun and the ocean were often depicted as a human metaphor, or as half human and animal, and clouds could have a human face to represent the wind, which was deemed a phenomenon of heaven. The activities of the spring season (i.e., represented in Pan and other deities), in part reflect the mythical imagery of the mind[35]. The pantheon of deities of Rome was adopted from Greece out of admiration for its civilization. These deities were not what every Roman worshiped at home, but they were a form of time-honored visual tradition. Roman design depicted human aspects in other ways too, that were not intended to be anthropomorphic, even though they resulted in this way. In some instances a structure can

---

[35] These mythical creatures have been recognized in psychology, in patient dreams.

show in its decoration a recollection that mirrors the mind, and the intuitive iconography of the Pantheon of Rome offers such an example. This recollection is achieved by its dome, oculus and coffering, and it is no small achievement to have arrived at such powerful suggestion by these means, in a design that resulted in a composition that mirrors the mind.

The Pantheon has evoked significant themes to those who contemplate it; the dome's hemisphere is a representation of the universe with the oculus as a somewhat literal analogy of the sun at the center of the dome. This opening is a practical design as it is too a source of natural light, in an otherwise closed hemisphere. It is a highly spiritual design as it lets the sun's strong light bathe the dark interior of the sphere from a high hole, and thus it produces an illuminated ellipsoidal spot that changes position as the sun passes through it. The composition of the Pantheon's dome is like a spherical planet bathed by the moving sun, except that this light reaches this planet's interior. It is a half sphere on a cylinder, which is as high as the radius. Hence, the design concept is that of a complete sphere expressed in usable space: a cylinder. The concave structure offers an enveloping space. The visual effect of this form was consciously driven by the connotations of geometric ideals since Rome applied Euclidean geometry to construction and the circle was seen as an ideal and perfect form. The Pantheon is a grand centralized interior that expresses a regularized organization of elements in a classical visual syntax, namely, symmetrical, proportional and equilibrated or leveled to gravity and the horizon. The solemn regularized pattern of coffering gives a mausoleum character to the monument. Among history's domes this is not a dome placed very high above. The position of this dome is nearer the visitor than that of later domes, and its impression can be that of a crypt to viewers, being different from the impression of a highly placed hemisphere. The Pantheon is actually a memorial built to worship and commemorate all classical gods, goddesses, heroes and honorable men. In the Renaissance Raphael Sanzio upon his death in 1520, requested to be buried in it. It also has the eighteenth and nineteenth century tombs of Victor Emanuel II, Umberto I and queen Margarita.

As mentioned above, this design additionally offers an intuitive design composition which was probably grasped at some perceptual level: the Pantheon's arrangement resembles the space of the mind and the oculus (eye) is an opening from within the world of the mind. This intuition could be a subconscious cerebral objective that guided the semblance of the final form, through the veiled geometric and celestial design ideals that are intended in it.

Interior coffering is an elegant treatment for a ceiling that adds a dignified or noble semblance to interiors as it has been widely perceived over the course of history. It is evident in the cross section of coffering (see Figs. 45 to 47) that it takes away some of the weight or mass of a ceiling, and this practical aim could have strengthened its regular application in classical ceilings. In fact several other features of the Pantheon's walls, such as its carved niches, reflect an intention to reduce mass in the structure. Coffering was already in use in flat ceilings in Athens, earlier than in Roman domes, such as that of the Acropolis' Propylea and the Parthenon. Rome used coffers in apses and niches as well as in vaults, and sometimes they were eight-sided rather than square, reflecting a tendency toward a spherical effect in the coffer itself, as that observed in the Basilica of Maxentius (306–312 AD), (see Fig. 47). If coffering is designed inside a spherical concavity as in a dome's interior, it intensifies the parallelism of this form with the anatomy of the head (expressed in the regularity of classicism rather than in an organic design form). This is probably a reminiscence that is approached in design without studious intention (see Figs. 48 a and b). The aesthetic basis of this dome[36] seems correlated with an intimate mold of the cortex. Several aspects of the cortex's surface organization are sometimes evoked in vaulted ceilings of different kinds and in different times and cultures (see Figs. 49, 50, 51 for example). The grounds for this intuitive composition arise in the forms observed in the Pantheon and its overall head-like proportion. Coffered domes are an architectural design that migrated through emulation[37] (born in admiration and inspiration) to many other buildings of history for several centuries, hence, the compositions of cosmological nature in it, and the depiction of elements of the mind's interior reappear in other domes, and the repetitive choice of the design indicates a collective acceptance of a spherical design that is sensed by different generations to be fitting (good) as a solemn form.

The ancient psyche, apparently not on purpose, arrived at the design of a dome's interior as a mold of its mind. Such a conception could be an internal 'knowledge' of the logic of shelter, a pure intuition, or what the 'inner' mind senses, even though the eyes don't easily see it. The interior surface of the skull is the mind's natural hard cover (structure), a fitting form that is brought to fruition in design by the guidance of intuition, at least in concave design cases, but in the guise of the immediately-perceivable design concepts mentioned above. The brain has multiple surface mounds and the structures that protect the cortex have complementary indented forms to those of the brain's surface (like coffering). A skull's interior also reflects these mounds. A cross section of the Pantheon's coffering shows a lobed circular form that is similar to a silhouette of the brain. The skull is an intimate formal precedent for a concave form or a half sphere. A design composition that represents something as significant as this natural protective structure, reaches beyond the temporary state of affairs of its own time. This indispensable human surface of the brain seems to be expressed in the design of some later head attire used in funerary sculpture where something similar to reversed coffering or reticulated mounds is also expressed (see Fig. 50). This head attire design suggests the brain itself rather than a concave mirror image of the brain. The idea may perhaps be expressed behind some other different designs of building tops (see for example, the Crown Palace in Prague, in Fig. 52 which also brought to mind the brain to the author upon seeing it).

The Pantheon's coffering was added in 300 AD, in a renovation of the interior by Severus, and this means that the natural analogy of the Pantheon, whose clarity is enhanced by coffering, was aided by the input of later minds, which seems to point at a collective (possibly) subliminal sense of the head's anatomy in the deep fields of human creativity that can become associated to an architectural sphere. Inside, the Pantheon has a smooth finish, lined with *precious* marbles[38], whereas on the exterior, its breakthrough and historic dome lies behind a rather prominent gabled portico which is a different, more ancient, and at the time of construction, a more common type of design.

There is no better explanation today, for an architectural impression of the brain that manifests itself in the compositions of different designs, than to see it as the product of a little known natural

---

[36] Aesthetics enfolds all the aspects of a design that makes it attractive, among which are distinction, spiritual and intellectual quality (in an ancient monument), and certainly, an unforgettable composition. There are entire chapters and books devoted to the subject of aesthetics in philosophy.

[37] The most common routes for the migration of motifs are trade activities, production of design in conquered lands and reproduction of valued examples and prototypes.

[38] Regardless of our varied perceptions about aesthetic materials, or the practice of coating in architecture, marble was a highly valued and aesthetic material of ancient Rome and of history.

or cultural catalyst of creativity and aesthetics. It is not possible to explain (comprehend) how the mind produces all its creative ideas, even some which have not been voiced or named, such as these depictions, and maybe without ever being seen directly[39]; but it is possible to identify these intimate forms in the outcomes of vaulted and domed design[40]. The representation of the brain in highly aesthetic interiors is perceived in this study as an internal 'vision' of the mind. This is a preliminary interpretation and by no means intended as an incontrovertible truth. If we prefer to entertain a possibility that a brain design is planned, we must then ask why there is no knowledge or written record of it. Or is it so complete and natural a design that there has been no need to refer to it in the past? This absence reflects that even if someone might have observed the analogy of a brain in domes before, there has been at least no verbal or linguistic reflection about it, in design or in texts.

In presenting these observations, we also must remark that decoration does not seem in this case, superfluous or unnecessary, and much less meaningless. It may sometimes achieve a composition of crucial significance such as the brain. In the Pantheon's example, we should notice that this kind of cognitive design outcome is not related to a ruler's intentions to represent power, as mentioned in our introduction, but rather, it arises in the intuition of the builders who created a significant design image. This image is presumably what is somehow sensed subliminally in a viewer's mind, though it is not actually noticed.

Contemporary authors have deplored that the Pantheon has a trabeated portico that hides the greatest dome of antiquity. The builders of the current Pantheon's dome (built under Hadrian AD 118–128) apparently didn't think that the portico was awkward enough to be removed, nor did those who renovated it under Severus and Caracalla in the 4rd century. To a contemporary perception, this combination of different structures might reflect little freedom in ancient builders to have made a good design decision: to remove the front portico. The previous Pantheon was likely rectangular and gabled (trabeated), in the Greek temple fashion. The original was built in 27 BC by Agrippa. Perhaps its image was highly regarded, and this regard of ancient design would not be surprising to the author of this study which pays attention to the permanency of certain forms in culture; but we also see another iconographical force in the combination of pedimented cornice and dome together. This second iconographical connotation of the Pantheon's dome is related to its exterior. It has a composition that may make the portico appear somehow 'fitting' at the front of a dome, and in this case a recollection of an important object that is also closely associated to the human head. Ancient perception was probably moved by yet another analogy that acquires integrity as a composition by the two different structures of dome and portico together: a helmet of Italy with a tiara or pediment-like design (a triangular peak like that of a gabled roof). The Pantheon commemorated helmet-bearing gods, goddesses and heroes. The helmets of heroes are seen in ancient classical sculptures, some of which were Roman reproductions of effigies of Greek gods and heroic warriors. These helmets show a pointed form at the front top, and this point is repetitive in several different types of helmets of Rome and history. Its peaked shape is mentioned today as 'tiara', but the tiara form may well parallel the pediment of a gabled roof, therefore this recollection, even if subliminal, could have made the pediment of the front portico look habitual and fitting at some cognitive level

---

[39] The possibility of nature-given ability in intelligence is however available.

[40] Lobed circular outlines were also produced in circular floor plans, although a connection to the brain's lobes in this case is not immediately evident, and if it should exist, it would be perhaps, veiled behind a choice of configuration that is shell-like.

(see Figs. 53 to 59). There are historical indications that the columns and pediment, although not of an obvious physiognomic appearance, were cognized as a frame that can surround the face[41]. The pedimented portico is a structural image that migrated abundantly to aedicule, altars, interiors and frames, and helmets should not be spared. The physiognomic grasp of this facade above the face, as in a helmet's top, must arise in the culture's adjustment to the perception of this structure[42] and was probably behind a ritualized (simplified) triangular pediment or tiara shape too. This might point to a cultural archetype in the configuration of a Greek temple portico[43] just judging from its widespread and endless application in time. The visual association of the portico and the face was perhaps related to the use of miniatures that emulated a temple front, created to house a sculpted head or bust within its space (see also comment in Section 12, paragraph 7).

Roman soldiers wore helmets of different designs according to rank and activity (see Figs. 57 to 59). Handmade helmets were not identical to each other (in particular in decoration) but they had similar typology. The helmets for warfare, the *galea*, were not decorated as those of ceremonies, or those used in gladiatorial contests. Officers used a Greco-Roman helmet that is the most reminiscent to the dome and its pointed triangle above the portico. This peaked frontal band is also observed in the helmets of centurions (recorded in A. Rich from extant friezes and sculptures of Rome). This piece became movable in some helmets, especially in later ages. The similarity of the dome to the helmet of antiquity is not discernible to most people but it has been suggested[44]. There could be a natural concept of a protective cover in the mind, a cognitive sheltering sense that is like a head's 'vault', or there could be a cultural archetype in the helmet[45]. The parallelism of architectural domes and this domical head gear shows a cognitive homology that could have roots in the sensory perception of their sheltering effect, and this is the intuitive iconography of the Pantheon. Though the Pantheon's front portico has a depth that doesn't match the thinness of the visor design of extant helmets this dimension in the building obeys the practicalities of this type of entryway, in particular if it was pre-existing. In other words, the scale it had before is the scale featured in the final composition that recycled earlier architecture. But it is important to understand that an intuitive iconography of a helmet's configuration is not a case of replication of exactly every detail, but it is a recollection or the most basic features of a visual theme. When image migrates from architecture to objects, it does so by the adaptation of building parts into non-functional ornamental (or ritualized) designs. These are executed in different dimensions and materials. The Attic helmet itself as that shown in Fig. 56, could have adopted a temple pediment outline, even with a temple's gorgon at the center, as a ritualized pediment of no depth. Some other much later architectural designs featuring dome and pediment resemble even more an Attic helmet with a gabled front form. See for instance the portico that Ventura Rodriguez designed several centuries later in a domed pavilion of the church Our Lady of *El Pilar* (in Fig. 60[46]). Additionally, there is a small structure in Vienna in the cathedral of St Mary's

---

[41] Jul. Poll. lv. 133. describes a mask with a wig and bonnet arranged in a pyramidal form on the top of the head, like the roof of a house, or the Greek letter A. Cited in Rich's Dictionary of Roman and Greek Antiquities, under SUPERFICIES, page 633, 1893.

[42] Akin to the images of the mind previously-referred here as 'forms in the clouds'.

[43] The idea that cultural forms develop into archetypes has been proposed by C.G. Jung.

[44] E.B. Smith 1950. Smith reflects that the ancient helmet was influential in the conception of ancient domes; and the same book quotes that Plutarch compared the dome of the Odeon in Athens and a celestial helmet

[45] The parallelism of domes and helmets is not always subconscious. It can be identified (though as a curious coincidence) in vernacular cupolas and helmets of Europe. (The author of this current study noticed this resemblance in the nineteen seventies, before coming across E. B. Smith's book "The Dome", for example).

[46] The particular portico of Ventura Rodriguez seems to have tried to create a Berninian altar form as in the

Church on the Bank which is very similar to a helmet. This form could have been directly inspired by a helmet. The current cathedral dates from the 14th and 15th centuries.[47]

The Roman builders could have certainly given an intended reminiscence of a helmet to the Pantheon, as a fitting design concept for a dome. In fact, the dome of the Pantheon was originally covered with bronze plates![48] But behind any possibility, and whether helmet[49] or dome is the generative form in this pair of cultural structures, it is important to identify in both helmet and dome, the built expression of a cognitive (and sensory) idea of a guard or protection above the head. This head protection concept or internal vision, that for the purpose of naming it might be called a 'mind guard' is an iconography that ties together two different designs. The analogy between the two artifacts is connected by a sensory idea of protection for the head in a hemispherical or concave form (or structure). This intuitive mind guard idea is a free concept, independent from the domain of the practical and respectable mechanical development of the dome.

The helmet was in ancient times an honorable object associated to the responsibility and sacrifice of combat and to its protective service to combatants. As the very ancient Villanovan helmet indicates (Fig. 62), cognitive associations of the helmet can evoke a wearer's head (his identity) by being placed on top of a cinerary urn, in a position of a hat or dome[50]. In fact, warriors of many cultures were buried with their armor and defensive weapons; which speaks of both the vital protective role of these objects in earlier periods, as well as the honor of wearing them. A helmet had spiritual connotations in ancient times, judging from the idea of celestial helmets, quoted in ancient literature[51], and the depiction of helmets with wings in deities, such as the winged helmet of Hermes or Mercury. A recollection of a user can be brought to mind by certain objects of personal employ. A case of a chair in a late Roman tomb—that will be discussed in a later section—also brings to mind a user. In several monumental buildings and gates of Europe, there are sculptures of empty pieces of armor[52] sometimes like a lifeless torso or head without a face. This depiction and its tumbled shields appear like fallen soldiers in an explicit and dramatic visual way. There were sculptures of empty armor in ancient Greece as well, as those that depicted the enemy defeated[53], such as in the frieze of the portico of the Temple of Athena at Pergamon. Over the centuries, the theme may also address the loss of a nation's own soldiers.

A later funerary helmet of Prussia, for ceremonial use, displays a screen with vertically-oriented bars and it still has a triangular recollection (a thin ritualized pediment form) above the face.

Cornaro Chapel. It resembles Bernini's chapel in its curving diadem form, though it has the neoclassical syntax of Ventura's time blended with the Berninian Baroque (The features and sculpture of the Cornaro chapel were widely emulated in Catholic altars).

[47] The helmets of uniformed forces of the Austrian empire, in the nineteenth century, strongly evoke cupolas with an apex, and make these soldiers and policemen appear to be costumed as architectural towers.

[48] This bronze coat was melted down by Urban VIII for the construction of Bernini's baldachin.

[49] Helmets are more ancient than structural domes, but domical small huts are also extremely ancient.

[50] Lewis Kausel, 1982.

[51] For example, see E. B. Smith's The Dome, 1950.

[52] One example is the *Puerta de Alcala* monument in Madrid (see this chapter's heading) and Vienna's monument at Heroe's Square. Other examples can be seen above classical cornices in Berlin's *Unten der Linden* avenue. Schinkel designed a second proposal for the Berlin Guard House with such *tropaeum*. There is also an ancient sculpture of Athena in Dresden that depicts her contemplating her helmet in reflection.

[53] The Tropaeum. See opening image of this chapter, Puerta de Alcala of Madrid.

The bars have incised vertical grooves that together with the domed helmet and its suggestion of a pediment, seem to once more evoke a subliminal architectural theme of a dome and a colonnaded portal, though this piece would have distorted 'columns'. The fact that this later helmet evokes domed and colonnaded architecture has seemed a good reason to mention it in this cognitive-based discussion[54]. In this helmet, the memory of fluted columns and a dome are probably influential behind the conception of the longitudinal grooves in the direction of fluting; thus, the bars become compositional analogs of the fluted columns of the Pantheon by the direction of the incised lines, their spacing, and association to pedimental and domed forms. This is a composition analogy that was in all likelihood inadvertently arrived at. Bars of this type may be common in the metal crafts. These particular bars are curved (see Fig. 63) which indicates that the mind that designed this type of helmet had forgotten or buried deeply any idea of columns. The distorted columnar bars may have surely offered functional purposes, as they commonly are a face protection in helmets (the *buffe*), they enhance ventilation, comfort and access to the mouth, i.e., to scratch or drink. A parenthesis needs to be made to mention also style and transformation. Stylistically, this curving and bulging or swelling of linear forms may be related to a fashion for distortion of linear elements that became common from about the late '1500s[55] to the '1700s, and lasted until the nineteenth century. For example the fluting of columns often had a twisting gesture in this period of time, and influenced many other linear designs. In the sixteenth century men's pants and sleeves also displayed a linear textile design and a bulged shape, and comparable forms appeared in wood, metal objects and ceramics, where the torsion of linear elements also could bulge. This migrating design paradigm has a dynamic intent that abundantly uses a twist that evokes a kinetic turn in an object and it seems enhanced through lines. Fluting-like forms were also varied into some convex versions of the design (rather than a concave design) in metal and stone objects (see explanation included before Fig. 64). A clearly linear and changing convex form became common in the mentioned period of time and it is still designed today (see Figs. 72 to 74).

The sheltering iconography of both dome and helmet, to the head and mind, may play a role in the importance of these entities to all people. Thus domed structures are probably not too amenable to become forgotten, even if they may be left temporarily out of public interest at some historical point. Their obsolescence in some periods of history is temporary, the statement being derived from the tenacity of the design over so many centuries. The artifacts observed here and their parallels indicate that the concept of a cover or protection, in particular above the intellect (helmet, curved grand ceiling and metaphorically even the firmament), are somehow homologous[56] to human intuition in an experiential way, as they are all covers above the head, and hence there is a sensory logic of parallelism among all of them.

The example of the Pantheon as representing a helmet also informs that structure (though arising in the mechanical solving of static forces) needs not be far from the figurative grammar of the deeper

---

[54] Important helmets have been highly decorative. The craft of embossing and engraving armor has utilized many types of motifs, magical sentences and creatures that were thought to protect in battle. Architectural elements should not be surprising in armor. Medieval arms show many parallels with gothic tracery and perpendicular gothic lines. There is a parade burgonet in the collection of the Fitzwilliam Museum in Cambridge, UK, from about 1545–55, that was probably executed at the Filippo Negroli workshop of Milan. It shows a lion face topped by edifices with arches. A burgonet is a light steel helmet of the 16th century that originated in Burgundy France.

[55] Linear elements advanced toward a bulge that is observed in the design of clothing and fluted furniture legs. It is present in the age of Mannerism and beyond, and may be related too, to the evolution of bulging forms in Elizabethan furniture posts and the later transformations of soup dishes into the fancier tureens.

[56] Lewis Kausel, 1982.

intellect. A spherical structure evokes a semblance of the head and also of a head cover, and the two basic structures, arcuation and trabeation have had such physiognomic identities as expressed in the Pantheon and the Greek temple facade. Placing the temple front next to a dome, permits the inner intuition to design a building that brings to mind the protective configuration of a helmet.

Figures 45 a and b: Coffering in slab ceilings of Greece shows the relative amount of mass removed by it. *Source: von Egle, original publication 1905.*

Figures 46 a and b: The Pantheon'c ceiling and its coffering. Amount removed shaded in black. *Source: Adapted from von Egle and modified.*

Each coffer is a hollowed space, like a miniature vault, and therefore it has the supportive properties of a little dome with square plan like a cloister vault.

Figure 47: Octagonal coffering in the barrel vault of the Basilica
of Maxentius, Rome.

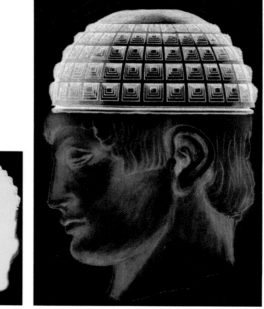

Figures 48 a and b: The Pantheon's interior space in silhouette (a) and in solid form (b).
The sightless vision of a dome to the inner mind. *Source: Original
artwork by Cecilia Lewis Kausel.*

Coffering produces a lobed outline as seen in Figure a, that evokes the outline of the brain.

Figure 49: Other brain analogy in design. Artwork by author shows the
    inner vision elicited by the Gloucester cloister vaults. These vaults were
    built in the thirteenth century. *Source: Original art work by C. Lewis Kausel,
    Design and Nature, Volume 3, 2007.*

Anatomical structures such as nerves, veins and some fanning out muscle projections of a skull are detected in the vault's design and decoration. The above design also shows souls of monks rising toward the center and thus forming a complementary image of a Grater Mind. This suggestion is more far-reaching than the vault's immediate structure. These instances of parallelism reveal the covering of the vulnerable head, where the creative mind and eyes are located.

Figure 50: The Pantheon's interior and a head attire with reticulated raised mounds.

Implied images of brain and skull. Composite artwork by author, part of Design an d Nature article, 2007. (From Art Exhibition, Mount Ida College, 2009). © Copyright Cecllia Lewis Kausel. 1982.

This head attire is observed in funerary sculpture of the 15th C. In the representation of a lady with a mantle of head mounds the portrayal seems to include her mind (perhaps its power and vitality). The wish to portray a person's body that has began a process of disintegration may compel funerary sculpture. In other words, the wish to remember the individual or a way to 'suspend' appearance as it was before death, responds to a necessary bereavement and tries to compensate the loss. The coffered or reticulated mantle mounds may seem just fitting for a funereal setting.

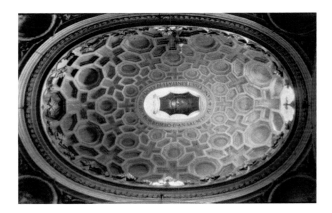

Figure 51: Oval Baroque dome with octagonal coffers may evoke a vision of a skull's interior. *San Carlo alle Quattro Fontane's* dome. Started in1834 by Francesco Borromini. Rome.

Figure 52: The shiny globules on top of this building may subliminally evoke a brain surface in this early twentieth century building, at least from the viewpoint of a passer by, but presented in aesthetic pearls.

The Crown Palace. Art Nouveau building in Prague. Built from a design by A. Pfeiffer.

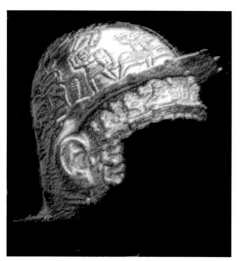

Figures 53, 54, 55 and 56: Helmets of Italy and the Pantheon without its front colonnade.

Top: Roman parade helmet of the late I AD to II AD (rendering by author after a metal reproduction by Deepeeka). The original was found in the 1700's near the Calvary Fort at Ribchester, England; British Museum. The front is a mask that is secured to the helmet by a tab.

Bottom: The same helmet excluding the mask and the Pantheon (right sketch) excluding the front colonnade, showing only the cylindrical body of the building and its rear cornice with pedimental design. *Figures 53 and 55 sketched by author.*

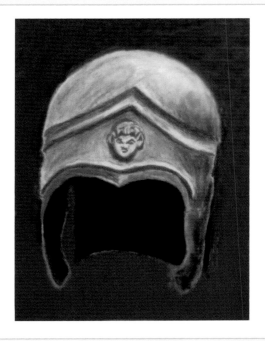

Below: Author's sketch from an original bronze helmet of Attic type, form southern Italy, with tiara (pediment design), likely the form that became a visor later. Central head motif is like a pediment's gorgon (author's view) which were said to have protected against the evil eye; 5th B.C. Drawn from internet photo 'attic 29' courtesy Legio VI's (copyrighted by Maical Förlag – GML Ex-Guttmann Collection).

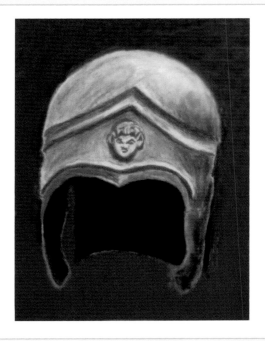

The attic helmet could have adopted a temple pediment outline, including a temple's gorgon at the center, as a ritualized pediment of no depth.

Right: Small structure in the cathedral of St Mary's Church on the Bank in Vienna. This form could have been directly inspired by a helmet. The current cathedral dates from the 14th and 15th centuries.

Figures 57, 58 and 59: Minerva's helmet, centurion helmet and soldier helmet.

In Rome, the tiara form often protrudes like a visor. The Minerva visor form with side scrolls is observable in Greek sculpture, as for example in a slab entitled Companions of Telephus in the Pergamon Altar.

(Continued)

Left, previous page: Helmet from a statue of Minerva. Roman officers wore helmets based on the Greek Attic style that remained popular with gentlemen for centuries. Second diagram: The same helmet's frontal view with ear pieces raised.

Right, previous page: Three *Galea*. Centurion helmets with a crest of feathers (the *galea* developed from the Montefortino helmet and before, from Celtic helmets). The *galea* helmet is found in Gaul, Spain and northern Italy. These three illustrations were made from the Arch of Trajan.

Bottom, previous page: Helmets of common soldiers portrayed in the column of Trajan. A. Rich. 1893.

Figure 60: The tiara form blends with a pediment with dentil decoration.

Domical front pavilion of the chapel, *Nuestra Señora del Pilar*. Ventura Rodriguez. Zaragoza, 1753. The architect created this design that parallels a helmet most likely without consciousness of such an inspiration.

This pavilion front may recall even more the Greco-Roman helmet type than earlier domes. This is a reference to its frontal design. The dome has openings that admit light that these figures don't show.

Figure 62: Once a personal protecting cover worn by a warrior, this helmet evokes the fact that this urn contains the ashes of the owner.

The helmet, rather than another object (i.e., a sword) recalls the head (maybe the mind) especially due to the way it covers and fits the urn. This position is also the situation of a dome to its space underneath.

Funerary urn of the pre-Roman Villanovan culture, a Roman settlement near Bologna. The Villanovan people inhabited what today is North Italy, centered at Bologna. They started migrating to Etruria from 100 B.C. onwards. This urn is probably pre-Etruscan. *Author's graphite drawing, 1982.*

The fluid twist and bulge in design

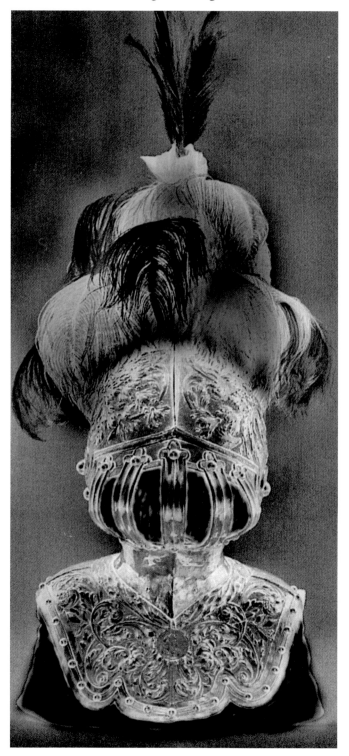

A dome with a peaked triangular front and colonnade is brought to mind by this helmet. The columns are recalled by the longitudinal lines incised on the distorted bars. The evocation is subtle. It was carried at the funeral processions of all male members of the royal house.

Figure 63: Funerary Helmet with triangular opening and distorted columnar forms. Prussian, 1688.

Dynamic transformation of various kinds of fluted and linear elements in late Renaissance design and beyond.

The twisting appearance and bulged forms that became popular in columns and other design.

We must leave out many additional things that can be said about each of these forms. Columns with linear and winding rope design are observed in substantially earlier art than the sixteenth century, thus this betrays an ancient origin for the idea presented here, however, the idea did not seem to occur before this date in ways that involved all the crafts. Ivory, stone and metal arts had used a rope design on rectilinear columns since at least 400 AD (in Syria, Europe and Asia, and Roman metal artifacts displayed similar designs in artifacts of Pompei.

Focusing on the technical basis for the twist, this is achieved by a rotating mechanism such as the potter's wheel, or a carpenter's turning table. More than one craft may have an activity where a twisting gesture of vertical elements can be achieved.

In architecture, we find theories that claimed to have discovered that the pillars of the Temple of Solomon had twisting columns, and these were incorporated by Bernini in his baldachin. The Baroque twisting columns of St. Peter's basilica were immensely influential for later Catholic altars. The image was of interest to the carpentry and furniture crafts as well. Solomonic columns are found in Baroque wooden entrances, stair balusters and chairs in England, for example. These columns appear energized (as in living forms), and they carry this life-like iconography. These columns are certainly later than the twist of rectilinear forms, but nevertheless they were influential in the popularity of the twisting gesture of the Baroque age.

It is of no less interest that linear designs became popular in clothes in the late Renaissance (before Bernini's columns), giving to textile parts such as trousers and some sleeves. a ribbon-like motion, or flow. While this fashion is different from the twisted columns, it seems that it could have a cognitive relationship to fluting (see Figs. provided). We must not overlook the subliminal place of the human figure as the center and purpose of design. Hence, the way clothes make the human form appear at this time, may not be without some power in promoting fluid linear bulging forms in the crafts, which could have derived from the ubiquitous flow of lines in the appearance of men's sleeves and trousers.

It may be very difficult to pinpoint which craft was the origin of the twist. If the source of a pattern does not originate in a static form, it can come from several possible stem forms. But the

Figure 64: Twisted bulging column. Painting shows a 'life-full' column.

The flowing fluted base of the thick column gives a visceral evocation to the representation. Jacopo Bertoja 'Entry into Jerusalem', 1568. This depiction of a column may follow the theoretical work of Diego de Sagredo of 1526.
  At the Oratorio Gonfalone, Rome.

twist is still a visual phenomenon that permeated all design. It probably began gradually, and the available mechanisms in the various crafts that could achieve the twist, provided a necessary ease of manufacture for the torsion that suggests motion and energy. We must always expect that the movers of popular forms are compelled by the cultural mind. In this affair ideas undergo mutual feedback, as the recognizable form is doable in several crafts.

Figure 65: Twisted fluting on rectilinear columns.
Source: Drawing detail from Warth.

While this particular twist in rectilinear supports was popular at this time, the idea is not clearly that of a living form yet. This twist was observable in the decoration of ancient Rome and in the arts of metal. The man's pants show the bulge and linear pattern acquired by men's clothes in the sixteenth century.

Left: Common clothes of soldiers and workers of the 16th century.

Second: Detail of Franz Pforr's *Kaiser Rudolph of Habsburg in Basel*, 1820.

Third and fourth: Clothing of mercenaries, from Meyers Lexicon.

Figures 66 a, b, c, and d: Men's clothes and uniforms of European mercenaries showed linear elements at the calves. The breeches were lose and sometimes flowing as in the work clothes shown in the left figure. The stockings make the flexible textile linear patterns bulge at mid leg calf. *Source: Meyer's Lexicon.*

Figure 67: Columns twist in a motion that evokes a life form, such as a muscular snake that sheds its skin.

Saint Peter's Baldachin by Gianlorenzo Bernini, 1624 to 1633. The bronze for these columns came from the Pantheon's exterior. It is considered the first truly Baroque design, though it is not the earliest. It blends architecture and sculpture.

The lower linear pattren evokes fluting, that has been twisted, and out of which a dynamically moving core comes out as in a living form that sheds its fluted column skin twice before its top most segment.

Figure 68: Monastery *La Cartuja*, Granada, Spain; Ciborium, by Hurtado Izquierdo 1727–1764. The columns of St. Peter's basilica acquire a smooth and wet look (through polishing) of more slender elements that brings to mind umbilical cords.

Figure 69: The highly polished twisted columns acquired a wet look (living-like) that resemble entrails, in particular umbilical cords.

Baroque altar details of Solomonic columns; Parish Church of St. Stanislaus, Poland.

Figure 70: Twisted columns recall umbilical cords and the sinuous surfaces look organic as in natural forms.

Detail of twisted Solomonic columns in the *Hofkirche at Wurtzburg's Residenz*. Balthasar Neumann, eighteenth century. Began in 1730. The palace chapel, designed by Neumann and decorated by Hildebrandt has a basic plan of intersecting ovals. Its interior architecture and decoration is dynamic, uses polished materials and transformation of the space so that walls flow into ceiling without a sharp break.

Figure 71: Twisted Salomonic columns are common in the historic buildings of Granada.

Figures 72: a, b (above) and c (left). Designs in silver, glass and stone originating in ancient patterns such as those of Pompeian metal and pottery objects.

This type of decoration became popular in the period of the twist and it has come back today in traditional artifacts. The twist of linear elements of Europe seems to show transmutation in its many interpretations, as certain design forms seem related to column fluting that changes its dimension as the object bulges.

Above, left and right: Contemporary reproduction in pure silver of a chalice from 1756, Greek. The linear design is convex rather than concave (as fluting is) and has an exaggerated bulge and ornate decorative pattern.

Left: Design by Karl Friedrich Schinkel for a glass cup from 1829. This twisting pattern changes its dimensions and it is visibly related to classical column fluting.

Figures 73: Design in stone originating from ancient
Pompeian patterns of metal and pottery objects.

Left: Garden vase in front of the greenhouse of the
Burggarten of Vienna, by Frederick Ohmann 1901. This urn's
twist design is also reminiscent of fluting.

Figures 74: a. and b. The twist of linear designs in the work of artisans. American.

Chapter 5

The Emulation of a Model: The Creation of an Architectural Motif

The human psyche is receptive to well-balanced looks even if they are non-figurative and analytical, as construction systems are. Evidently, structure by itself sometimes has a sculptural quality, and builders can enhance this appeal in finely crafting a building and its details. As indicated by the ultimate reuse of structural images as embellishment, it is clear that the forms of many ancient structures are appealing to the public despite their conception as static systems, rather than as ritual or ornamental forms. Some historical utilitarian constructions have been endowed with very minimal or unobtrusive visual effects that can make an interested viewer who is admiring its intricate design, observe also its harmonies. We can notice that this was done in ancient utilitarian public works such as bridges and aqueducts (see *Aqua Claudius* in Fig. 75, the Aqueduct of Segovia in Fig. 44 and the aqueduct *Los Milagros* in Fig. 125). But ornamental treatment was certainly more marked in architecture while it is frequently almost indiscernible in public works. The Porta Maggiore of Aqua Claudius, however, was ornamented between its arcades, with post and lintel systems with pediments that are unnecessary in the support of the structure. They must be seen as embellishments of the piers. The design of structures, once built, is often understood in perceptual terms as an image, as is clearly observed in these added temple fronts of *Aqua Claudius*. Therefore, essential supportive forms, not unlike ritual or ornamental design, are perceived conceptually. For example the image of the aqueduct of Segovia is used in miniaturized carved form in gravestones near the aqueduct[57]. Consequently, structural systems did (and do) offer images, both to the masons and to culture, to become inspired by and to use as adornment. For sure some cities are best known because of an exceptional structure sometimes, such as the tower of Pisa for example, and this causes local artisans to depict the tower in crafts.

---

[57] They are left unexplained by Casado, our source. However they could arise in a wish to give dignified burial to people (who could have perished at the site).

A building's static support determines most significantly the overall shape of its body, but most changes in style have been achieved by the ornamental treatment of architecture. The structural types of history are surely less numerous than the styles of history. A simple example can be explanatory. The temples of Egypt, the Palace of Knossos and the temples of ancient Greece, all used a trabeated (post-and-lintel) support. So, they all resorted to halls of columns that support stone lintels. However, their ornamental syntaxes (treatments and proportioning) are done in different styles. Comparatively, semicircular arches have been a main structure in classical Rome and the Romanesque, but the two developments are stylistically different. A purely ornamental style that does not result from the evolution of structural form probably arises from smaller technical arts and customs as well as other socio-cultural variables, whereas the changes in structural support (which also bring about change of form, hence, style) are a function of the discovery of new and more advanced static systems.

Another observable effect of the culture of structural motifs is the blending of an older structural form with a new structural system. This is a manifestation of design continuity, or a case where a no longer functional form is represented[58]. This is possibly decided for the intent of conveying a well-known visual character over a newly-exploited structure. When structures changed or advanced in history, the newly-devised supportive systems had a resulting form that was naturally different from an earlier one. These achievements offered opportunities to change building appearance and introduce a radically different architecture; opportunities that would have been the dream of the twentieth century modernists, who looked for ways to produce different architecture. However, the evidence of ancient times shows that care was taken to make smooth transitions of building appearance[59], at least this is the case in the classical monumental structures that have survived. At such times the record shows that masons and patrons (some of them possibly acquainted with the aesthetic theories of Aristotle) willed to depict elements of an older style over a new structural system. In other words, they fashioned design continuity by adding certain elements of a previous system. And again, in the tendency of society to make new forms that were visually related to the old, a newly-introduced support was actually embellished with the looks of the previous one. Next we illustrate the case in point. Thus, the ancient cultural psyche seems to have avoided a drastically-different novel appearance in monuments, not to mention that it also avoided plain appearance in common urban features as in water fountains and benches next to walls, endowing these smaller structures with faux pediment and columns too[60].

Rome exploited the true arch when masons understood that it is a much stronger, more efficient structure than the trabeated system[61]. The arcuated structures that had existed for long were understood at this time in a way that can be considered an engineering event more than an

---

[58] In the aesthetic theory of ancient Greece, Aristotle and Plato referred to all art as a case of mimesis (imitation) and it was used in the sense of a 'representation' rather than a 'copy'. The term 'mimicry' has been used to refer to animal and plant resemblance for adaptation that provides camouflage, protection from predators and other advantages.

[59] Similarly, the transition from Early Christian architecture to later styles, or from Romanesque to Gothic, always maintained a number of familiar elements and thus composition and syntax (regardless of change of style) was also maintained between two consecutive developments.

[60] This visual objective seems to account for structurally unnecessary post-and-lintel forms over the piers of *Aqua Claudius* as well, since aqueduct builders must have understood the strength of the arches of their structures well.

[61] The arch, before this historical moment, had an ancient application in Egypt, Assyria and China.

architectural one, even though the semicircular arch is a very simple case of engineering. The use of these structures in public works, and later, the Roman experimenting with arcuated forms in large scale structures, has given the rationale of this thought. At the time of exploiting the arch in triumphal arches, the pre-existing trabeated structural system was blended with monumental arches (see Figs. 76 and 77). Trabeation is not needed for the arch to function. The application of trabeation to structural arches indicates the continuity of a trabeated structure as a motif. It is interesting to ponder deeper on the reasons for maintaining similarity between two structural styles, especially now when we have gone through the twentieth century modernist design value of seceding with tradition. One such reason could be that society identified a well-established status[62] (at that time) in the monumental trabeated system[63]. For example, the post-and-lintel system was in use to feature a seated public statue of a goddess or an emperor. Possibly, the new idea of a triumphal arch should be no less ceremonial than these previous examples.

This case already showed in history the common proclivity to continue favorite systems even after an old method is displaced by a new and more efficient one. This effect may result from a combination of factors. First, it is a trait of culture to maintain some designs. Second, in ancient Rome, the aesthetic effect of Greek trabeation must have offered a visual theme, a known configuration to the lay mind in particular, to readily relate to a new monument[64]. By 'theme', we also mean to say that the trabeated system had become a collectively-grasped motif that had become satisfactory and meaningful to the society. The trabeation of an arch linked the structure of the arch to the classicism of Greece and its civilizing connotations, which Rome had fully embraced. Third, presumably, Rome's rulers were additionally interested in embellishing the plain arch so as to give it an ornate and stately classical character. It has been mentioned abundantly that to those in authority in Rome, it was important to give impressive monuments to the masses. The patrons of triumphal arches probably thought that a grand and imperial classical look rested in Greek trabeation. In regard to the builders' preference in these matters as opposed to that of rulers (if it was in any way different), the architectural literature has mentioned that their inventiveness may have been only auxiliary to the wishes of those in authority. But at any rate, there were Roman masons who also regarded the trabeated looks of ancient Greece as a refined and tested standard of excellence. This reverence for Greek design we find recorded in Vitruvius' treatise[65].

If the arch is flanked by massive enough piers or walls there is no need for the lateral abutment of trabeation (see Fig. 76 lower left). Thick enough walls (or piers) carry the lateral thrusts of the semicircular arch on right and left sides. The columns shown in the triumphal arch (Fig. 77)

---

[62] It was crystallized in the minds of those who decided.

[63] Whereas the arch was known in public works, prior to its implementation in triumphal arches, there was no stately or ceremonial design associated to it in classical Europe.

[64] A theme permits people to readily identify at least the thematic aspects of any newly-created art. This view may not be the only reason for themes to exist but one that is certainly reflected in studying this topic. Themes aid recognition in all the arts (fine arts, literature, music and poetry), and hence, the public upon first seeing new creations, is helped by the presence of a theme, to identify what the art or design is about. In architecture, a theme is not a story but instead something like a visible type that is discernible when it is reinterpreted in newer buildings. The public is previously acquainted with the type's elements.

[65] Vitruvius was active in the I century BC and wrote his treatise to Augustus. Later writers have implied that Vitruvius wrote about Greek rather than Roman architecture. This perception is influenced by our contemporary ways to rank design by function, and it addresses the fact that Vitruvius' admiration focused on the stylistic forms of Greece rather than on the superior structure of the arch.

do carry downward weight from the top cornice but this is not an element needed by the arch. It results from the added massive trabeation. Thus the thick walls (or piers) provide the lateral buttressing needed by the arch rather than the superficial columns. Anyone may still think that the age of the early builders could have solved intuitively, or by trial and error the thickness of piers in the triumphal arch, and not necessarily with a clear understanding of relationships, therefore they added trabeation with the intent of helping the monument stand. Despite this type of possibility, it can be firmly assumed that many ancient masons discerned that the arch didn't need the trabeated structure because plain arches had been in use in public works (and indeed in the rest of the arcades of Aqua Claudius), and were known for a long time in the Far and Near East, and in Greece and Etruria. The trabeated façade had had visual power over the Greek psyche and it was to have the same effect over that of Rome. Several factors make it very likely that the Greek trabeated order was applied over the Roman arch to give a desired classical character to the monumental arch, which at the time of its introduction in buildings, it did not yet have. The example shows us the conscientious incorporation of a structural form when it was not necessary since the arch was a better static system. And as noted in the architectural literature (e.i., John Summerson), this triumphal order was achieved in Rome without hiding or masking the structural arch. The finely-sculpted identity of classical Greek design wasn't left behind in Rome, and thus, there was a visual fixation with the Greek temple façade. This temple front was continued as a motif and was never completely left aside being perpetuated still today.

The application of Greek trabeation over arches surely became less rigorous in time and after the Byzantine age the arcuated forms clearly dominated. In the Romanesque and Gothic ages columns were transformed substantially. They are medieval piers that are sculpted as bundled-up slim columns for a look of slenderness in massive shafts. These piers receive the thrusts from the vaults. Eventually, builders, but also patrons, must have come to realize the superior structural strength of the arch, but perhaps more importantly, they appreciated its roundedness as fitting and pleasing. This cognitive esteem is shown both in the many ways the arch was used and in the sculptural treatment it received. Though impossible to tell, probably when the appreciation or enjoyment of the arch as an aesthetic form developed collectively, it stood as a regarded image in the Western culture, becoming another design pattern of destiny that is used to communicate visually the fitting arcuated forms. In the Middle Ages, stone carvers came to design a sheltering effect in vaults and miniatures by means of the arch's circular nature. By that time, the arch's aesthetics was 'loudly' expressed in buildings in an implicit symphony of rhythmically multiplied arched moldings (see Fig. 138).

Figure 75: Classical trabeation as a motif in association with functional arches in Aqua Claudius aqueduct, 52 B.C. Began by Caligula. The aqueduct's *Porta Maggiore* has decorative Greek trabeation at the piers. *Source: Image from Meyers Lexicon., 1902.*

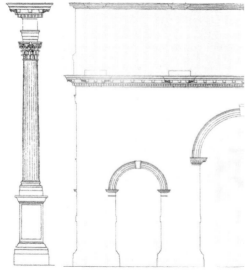

Figures 76 and 77: In the Roman monumental order the Greek post and lintel image provided status and a classical theme to the arch. *Source: Image from von Egle, 1905.*

Top left: The Classical Corinthian column.

Middle: The arch without columns.

Left: Triumphal arch of Septimus Severus. The Roman monumental character arises from the combination of arch and the ancient Greek trabeated system (see Summerson, 1966.) The addition of the trabeated system neither contradicts, nor masks, the function of the arch.

Chapter 6

Supports: A Medium that the Mind Transforms and Shrouds in Mystery

Enrichment with various classes of imagery is an ancient idea that was consistently practiced side by side the utilitarian body of a building. As a supportive element, a pillar matters highly and perhaps due to its vital role, it has been abundantly a medium for representation and to express a phenomenon of alteration and blending of different forms. A high variety of imagery has been utilized in columns and posts to create some other impressions in them. It has been important to historical mankind that architectural supports are treated aesthetically. One of such type of impressions has been the mysterious hybrid forms that appear in pedestals and furniture that can stimulate interest in the art collector, or the buyer of rare pieces and curiosities. The art-inclined mind (rather than the investment-focused, if we can make this distinction for our objectives) pays attention to the design of special pieces. For the sake of our study's search, let us think of the mystery that attracts in these pieces as that which the mind intuits as having substance but cannot visualize with clarity what it is, and let's now move back in time to the possible purposes of depicting animal and mythological forms in building support.

The ancient imagination merged animals, humans and architecture together, and this is most often expressed in pilasters and pedestals, but also in brackets, cornices and other architectural parts, and in gardens. Beliefs about the protection of graves by some hybrid forms travelled with people, and were learned by others, becoming customs that lasted in history. At the time of the Middle Ages we find an evolved and changed hybrid imagery, the zoomorphic decoration that shows formal points of comparisons with the earlier vernacular art of Eastern, Scythian and Celtic-Saxon cultures. We additionally see at this time, the representation of the animal persona of saints, in particular, in the lamb, the dove, the lion, the eagle and the ox, which are theriomorphic forms of Christ, the Holy Spirit and the evangelists. The European Middle Ages continued to depict griffins, chimeras and satyrs as the ancients did before them, and added monkeys, frogs, dogs, boars and monsters; but the medieval mind was concerned with discerning good from evil, and to teach monotheism, while these hybrids had been forms used by pagans; thus many theriomorphic ancient designs became destined to the representation of vice and demons. But despite the religious trends of the age, superstition was strong, and monsters were used on the roof of cathedrals in a protective role, to guard from invisible evil, or they could be depicted under the feet of incorruptible saints, whom devils couldn't tempt. The study of ancient societies reports that the historical mind was both fascinated and repulsed by the evidence of similarities between animals and mankind. These two perceptions can be hinted in the hybrid grotesques of architecture. Later, the Italian Renaissance rediscovered the classical hybrids, and approached them with more neutrality, though not all carvers did so, and we continued to see in some pieces the face of the devil in the later satyr.

The Renaissance opened the world's eyes to a high art that focused on the revival and cultivation of ancient classicism, where the ancient hybrid forms were one of its motifs[66]. However, during this and later periods, design elaboration became a means to offer an impression of magnificence and wealth[67], and we observe an abundance of classical ornament in the forms of cherubs, masks, animals and hybrids, all utilized by the human desire for lavish possessions. Society has cherished the high expression of the Renaissance and the ancient hybrid forms thrived without disappearing, being still found today (for visual reasons, and some unrelated to wealth, but perhaps related to our subconscious), in traditional Italian-styled interiors and furniture. We proceed next to a discussion of some cognitive aspects that are related to a possible perception of supportive roles in animal and hybrid forms.

Human, animal, and hybrid creatures were depicted in pilasters, pedestals and walls in more than one ancient society, and their representation in supports, deserves a little attention, even though the origin of many imaginary creatures is fantastic and more ancient than architecture itself. The blending of animals and humans is observable in early cave art, for example. The herm proper, the man-pedestal, was a stone marker of ancient Greece[68] which was thought to give luck to travelers, hence, it was associated to a belief that this stone slab depicting the god Hermes protected from adversity. Animal pillars are found in the East, such as in Mesopotamia and India; and are certainly observable in China, where pillars show dragons and sometimes birds around a column shaft. Our study's purpose is not exactly the study of myth, though if it can elucidate something, it deserves our attention. Our search is for an understanding of the role of intuition, imagination and sensory perception in the conception of design, and to investigate the pertinence of imaginary forms in architectural support, i.e., the imaginary roots of supportive parts.

Just as woven patterns of very early rugs and wall mats were reapplied in stone carvings, at the time stone buildings were built; animal forms may have been portrayed because of a possible ancient association of animals to buildings, plus the tendency of ancient societies toward mimesis of pre-existing design. The religious architecture of India has reflected a highly meaningful use of animals in supports, and this has been related to religion, hence, this subject

---

[66] The school of Raphael excavated ancient Roman houses in about 1599, such as the Golden House of Nero, and found these forms in grottoes. The hybrid forms were revived in design in the Renaissance and referred to as *Grotteschi* (grotesque). The ancient name in Latin was *telamones* (A. Rich).
[67] We bring to mind the Medici patron, Lorenzo the Magnificent, who permitted the best Renaissance talent to fully flourish.
[68] See the caption of Figure 104 for the origin of herms.

is clearly complex by involving beliefs and philosophies. However, in cognitive and sensory terms, the forms of supports seem to be repetitively a design-and-mind medium, and there needs to be a starting attempt in understanding the import of this imagery of living forms in structures, hence we continue with our analysis. Unlike bundles of reeds and woven wicker, life forms are desired semblances[69] for architectural parts, rather than an aspect of techniques. The hybridized pedestal has been used abundantly, and it reflects a revealing metamorphosis, and sometimes distortions of its inherent form. It also seems to reflect a possible tendency of the ancient psyche to associate supportive elements to life energy.

Life forms imbedded in a building or piece of furniture could have made sense[70] back in time. In order to try to understand if animals could be at all significant as cognitive supports, as the walls of Babylon, Persepolis and the porticos of Tall Halaf in Syria seem to show (see Fig. 84). People certainly mounted horses, camels and elephants, and from here, their legs could have passed to Assyrian thrones. We should perhaps begin by first picturing in our mind a natural world with no urban development and only animals, plants, rocks and human imagination —as the sole early resources to build. Then ask ourselves, how does any mind begin to conceive some form of sheltering system from the available elements? Does it start by visualizing spaces somehow supported by branches or other materials, or animals, or themes seen in dreams? Our imagination brings to mind the natural cave, since early human findings occur in these spaces, and also the tree, because our eyes exposed to post-and-lintel architecture, see a post's resemblance to a tree trunk. We can certainly trace plants, especially palms and lotuses in the shaft of columns of Asian and Eastern conception, and the Egyptian papyrus is carved in columns; however, the depiction of anthropomorphic and theriomorphic forms shows a very widespread application of another type, in walls, buttresses, columns, gates and stairs; that at least overtly, seems to have been favored as suitable imagery for supports. Why is this so?

Animals and hybrid human-animals, such as winged bulls with human head, satyrs, mermaids, sphinxes, fauns, chimeras, appear repeatedly sculpted in supportive elements, and the connection of these forms with support doesn't die out in art or in furniture. These reasons are strong enough for a cognitive investigation of their long application. There are functional and sensory logics in certain forms, as in the human figure, which can appear in supports both intuitively and ritually. Such imagery may not have thrived in all human groups however, because different cultures created their crafts focusing on different priorities, and while some ancient cultures were figure-driven others were not. Some regions did not have an anthropomorphic pantheon of gods, and others came to avoid human and animal representation out of faith-based respect. But life forms in pillars are suggested in very distant ancient world regions where diffusion seems less likely than an internal logic that may arise in pure cognitive or emotional reasons, such as the holding, and the bearing functions of the body, and the qualities of animals considered admirable or desirable.

---

[69] A griffin for example reflects strength, agility and vigilance, and it was much depicted in tombs to guard its remains (A. Rich).

[70] A thought of using the herm as caricature comes to mind, in particular in Rome that sometimes depicted philosophers and famous personalities as hybrid creatures. Herms are a suitable outlet to portray a person in a hidden way, and to show the 'other' side of people (see Fig. 106). But these caricatures are additional to the connection of hybrid figures and support.

First, with regard to the choice of human figures in columns and other types of support, there are certainly functional relationships that can be cited in these portrayals; and not only sensory-functional correspondences, but also columns shaped as women and male satyrs. To start with, we might imagine that ancient ceremonies used honorific canopies held by individuals in some celebrations, for the purpose of carrying a ruler, or a ritual object; or to protect them from rain or give them shade. One of the functions of the human body in history was to carry items and bear weight. The bearing function was practiced everywhere, and on daily bases in ancient times. People carried water, produce, baskets and buckets with construction materials. They often hanged the heavier loads from a pole that was placed on more than one human shoulder thus creating a moving trabeated arrangement. There are many depictions in carved stone and paintings showing this custom. We still see a formation of a human group carrying together some weight or an artifact in a procession, for honorific or ceremonial purposes. Pallbearers still perform this carrying style to show caring and respect for a dead relative or friend. Pillars are probably a crystallized design of these formations of carriers, of something like a canopy or palanquin, hence the anthropo-mimesis of pillars. The human-pillar form was therefore naturally derived from cognitive exposure[71]; hence, pillars and columns can go back to either a pole that is carried or guarded by sentinels or to the human form of the sentinels themselves (see Figs. 78 to 83). The human form in a post must have crossed the imagination of ancient builders; however, a sculpture of the human form in each and every post, would not have been too workable a representation (even if artistic expression wasn't perceived in today's economic terms). Ancient Greece left us the exceptional caryatids[72] in the *Erechteum* (Fig. 81) and the names of some architectural and furniture parts reflect a cognitive bio-mimesis, such as capital, wings, monopodium (single leg) and tripod (three-legged), and these analogies are observed in different languages.

The many theriomorphic permutations seen in columns (the grotesques) that can be observed in architecture (and furniture supports) reflect fanciful insights of a designer who senses some kind of different substance behind supportive elements. We believe that this different 'soul' of supports is important for architecture. A representation may show an idea or wish of a mind, and sometimes even its fears, in addition to collectively accepted analogies of guardianship, strength, protection and even the presence of an invincible leader in a building. The practical representation of some hybrids may sometimes indicate the shoring up of a great load (see Figs. 41, 88 and 111). Early texts of history contain stories of amazing (and legendary) human strength and contests against feats of prowess, among which is the scriptures' chronicle of Samson holding two columns until making a building collapse[73]. Furthermore, in religion columns sometimes were given common names, as mentioned in the Bible in reference to those of the Temple of Solomon.

---

[71] This is not the only instance of the crystallization of a behavior. In fine arts there are others, such as the praying position of the hands the crowning ceremony, or the supplicant position of the body, that has expression in art and actual religious rites.

[72] The caryatids of the ancient Acropolis are explained with a story of matrons (a word that betrays an origin in the idea of mother) who were assigned to support the roof of this structure 'as a punishment', according to the Roman architect writer Vitruvius. There are other instances of 'punishment' type of justifications in history's texts which seem to reveal a search for a validation for something that seemed to a writer to have sacrificed a caste of people, or, in this case, for its mere representation. The caryatids are artistic and stand at a very visible point of the Acropolis. To the inner mind they may be one of the symbolic manifestations of strength. See footnote 76 as well.

[73] Atlas additionally carried a globe on his back that represented the universe.

Transformation between living forms and pedestals is often implied in hybrid sculptures. The idea of transformation in design has seemed informative to this study in terms of what has been hinted by a carver behind a support. Transformation appears in a much later Baroque sculpture as well, where plaster cherubs coat golden pillars. This is *El Transparente*, where a plaster cover is depicted as peeling off from columns and letting golden fluted shafts appear from within (see Fig. 88). This particular design depicts transformation in columns toward a gold (possibly connoting a superior and heavenly) skeletal material, though it doesn't reveal life forms inside the columns' interior, but just informs that life forms are a surface coat (perhaps perishable) and inside there are golden fluted structures (a divine or unalterable strength or support?) Themes of transformation are detectable in the Mannerist carved supports and walls depicted by Vredeman de Vries, Ditterling and others (see Fig. 107). These visions seem to say, in this and other cognitive manifestations, that the artist who had this idea felt compelled to carve themes of his imagination that the pillars suggested, different from the austere material of pillars and walls. The Mannerist and Baroque artists further played on the imagination of viewers, suggesting paradoxes and illusions in addition to transformation in decoration. Transformation is certainly a phenomenon of nature that is slowly affecting all life (and also inanimate nature) all the time. The psyche may well detect it and when transformation is depicted in design it could be interpreted as insight for what goes on in the world, as in 'immortality'[74] through renewal of life, or the renewal of perishable beauty in spring and birth, as well as the relatedness of animal and human life.

There is an enigmatic vision of a human-formed pillar that has appeared in design, namely, the interpretation that it is a goddess (see Fig. 83). This is also subtly suggested in the Greek column as well[75] and perhaps in the church of Our Lady of *El Pilar* of Saragossa[76]. Again, in this case we seem to encounter the Mother archetype in architecture. The name of the ancient site of the Temple of Hathor is *Ta-ynt-netert*, translated as "she the Goddess pillar[77]". This vision of the column as a goddess is numinous, certainly focusing primarily on those aspects of imagination that belong in faith and not clearly in the crystallization of a bearing behavior, unless this vision is deeper. A goddess is often a protective being, and this association might develop an indirect link between faith and a column's strength and permanency.

The vision of human figures in the recesses of temple walls, however, does reflect a correlation with dreams in which we encounter persons in discrete building parts. In dreams we may see acquaintances within specific rooms and sometimes even in wall spaces, and we may see some in an attic and relatives from our past in a basement. The dream is symbolic of personal situations, and the different levels may simply indicate that others live at a place different from ours, or that once were near and now they are far or gone. The visions of dreams however, may have influenced some features of interiors in ancient times, at least in some temples, as dreams were prophetic. There are a lot of images sculpted within walls in Christian churches and Asian

---

[74] Survival.
[75] See Chapter 11, paragraph 5.
[76] The pillar metaphor is used in Christian litanies to the Mother of God. The Saragossa church, *Nuestra Señora del Pilar* is associated to a cult to an apparition of Mary to St. James on top of a pillar, that is thought to have taken place in the year 40 AD, for the purpose of Mary asking James to build a church in Spain. From this apparition, derives the name *Pilar* for women.
[77] Internet site of this temple.

temples (e.g., in India) that could be connected to visions of the inner psyche that visualizes non-living persons (saints) as within the spaces of architecture[78].

Free-standing human forms can be observed in the Tall Halaf portico of the first millennium BC. These human figures are columnar in their placement and elongation, something that has been achieved by elongated head attires or in different similar examples, by elongation of the body, and they also seem to inform that the strength of a lion and a bull (or the concept they represent) are underneath that of the priests. Whereas in the Tall Halaf portico, a priestly cast of men metaphorically seems to resort to a functional logic when they support a temple roof, and stand on the backs of other supports (lions and bulls which may represent ancient cities[79], tribes, or institutions, or other cultural analogy), the European designer of millennia later, portrays a human physique or *telamones* that is literally holding a building, and we see in some examples the effort that the hybrid, pedestal-torso, makes to hold a beam. They are depicted holding a building cornice with their backs, necks and hands (see Figs. 111 and 115). Human formed *telamones* were depicted in ancient Rome long before their cultivation in northern continental Europe and England.[80]

Strength is also symbolized in the robust lion paw that can be seen in cabinets and pedestals. This paw or a lion leg has been many times depicted under a human torso. Indeed, lions in particular, are extremely repetitive in architecture, and they can be associated to protection[81] and certainly to leadership. Bulls are reported to have been often the animal forms of fertility gods. However, an ox is doubtlessly a beast of work in an ancient society hence, a source of energy. Other types of animals once carried a small palanquin or chaise-a-porter as those seen in some Middle Eastern caravans. Ancient stone architecture could have therefore recalled a vision of animal support in depicting them in static architectural parts. Such animals ought to be elephants, oxen, camels and horses. And there is a depiction of an elephant supporting a small obelisk in the Piazza Minerva behind the Pantheon in Rome (see Fig. 41), and some rare Roman friezes show elephants carrying small towers on their backs. There are other examples, as a French ormolu and ceramic figurine in the Victoria and Albert museum, of a rhinoceros carrying a Rococo clock that is larger than the animal itself[82]; and there is a very ancient vessel stand supported by an ibex at the Metropolitan Museum, from 2800 to 2350 BC. Figures in silver of an elephant with obelisk can be seen as table ornaments in the Baroque. However, the most tenaciously perpetuated hybrid motif is actually the lion foot that extends at least from ancient Assyrian times to today.

Cabinets and tables can show structural motifs with the supportive vertical divisions portraying hybrid forms. These images may inform some circumstances of ancient architectural parts. A pedestal that shows a change of form along its length, i.e., from column into a creature, shows us the transformation that we have already addressed. The idea of transformation has sometimes been further enhanced by distortion in some furniture, and

---

[78] When dreaming, architecture can symbolize our mind's domain. Its stored memory can be seen in its spaces and floor levels.

[79] For example Ninive has been seen as a lion in biblical literature.

[80] Hinduism sees Visnu as the god of the pillar of the universe.

[81] As in Feng Shui.

[82] The museum reports that when rhinoceros were brought to Europe, they became a focus of attention and were depicted in accessories. A rhinoceros certainly looks as a strong animal.

appears as a fluid metamorphosis of one life form into another, or of architecture into a pedestal. There are many furniture legs and chest supports —especially of seventeenth and eighteenth century vintage— that shift from the plumb line into dynamic bulging corners that inform fluid stretching and transmutation, sometimes suggesting the bearing of weight. Figs. 105, 106 and 110 provide extant examples of supports with life-like forms and from different schools. Some of these supports can display very long necks, torsos, or fish tails[83].

The blending of life forms to inanimate building elements has appeared bizarre to our age. The reason for this perception must arise in our inability to see consciously the ancient visions of architecture, such as the reasons for life forms in walls, or the men with bull bodies of ancient Assyria and Persepolis, or the Tall Halaf caryatids standing on lions, and other examples. But in the application of life forms the suggestion of function is possibly present in those hybrids that represent strength or effort. Many herms have no arms, just as they have a pedestal for feet, thus telling by this streamlining that they are more pillars than human sculptures: living-formed ones, to be precise. The historical psyche informs consistently in these designs that pillars stand for life forms and for their interchange or transmutation with human-like forms. The linguistic perception of the saying 'columns never sleep', also reflects an association of structures to a living activity, which implies too that a loaded support never gets a break or a minimum of rest, as it never ceases doing what it does. This brings us to the recognition of the service of a pillar or pier under a building load. The representation of a patient and smiling anthropo-mimetic weight bearer as that shown in Fig 111, is also revealing in this sense. The hoofed feet of this male evokes the satyr, which sometimes may have just meant to the sculptor the strength of an animal under the human form.

Imagery in design is a visual language that brings us into the important domains of iconography, aesthetics and bio mimesis in design. In iconography, the lion became fitting to the crafts of architecture and furniture, reaching a wide application and long continuity. Animal forms can suggest desirable attributes in design, that are well exemplified in the griffin, for example, a hybrid creature that informs an eagle's vision in a lion's body[84], possibly standing for speed and swiftness. A griffin may display a frightening territorial expression too; hence, it is a guardian with the particular territorial 'skills' needed to guard a grave. In the lion foot of furniture, the suggestion might have been (in addition to royalty) a cat's equilibrium and natural weapons (as there are some designs of heraldic type that feature claws, or the ball and claw foot of furniture); however, a cat's features would also be chosen because of the undeniable human interest in these animals. The lion as support on the other hand, did make some kind of practical or other sense to ancient builders, even if its application bordered fantasy.

Before identifying the architectural associations of lion figures that relate to building functions, we will briefly acknowledge a few depictions of lions and other animals in buildings that do not seem to have a supportive logic, or at least are not directly associated to it, so as to account for a minimum of the animal data in an orderly way and be clear too,

---

[83] We must acknowledge also that secular design caters to the appetite for hedonistic design in users– in particular after the Middle Ages. It is undeniable that design is an application that makes life pleasurable. But the permutation of creatures and pedestals is probably more than the attraction of curious forms. It appears in Hiberno-Saxon, Scythian, Asian, Native American, African, Middle Eastern and other designs, as well as in folk tales (the frog-prince, the man-bat or vampire, and the man-wolf). Transformation is as an attempt of creativity to communicate notions.
[84] A. Rich.

about the significance of animals that is external to the conception of static elements, but certainly not divorced from the aesthetic domain of architecture.

Architecture is a medium to carve scenes, thus many depictions of animals are meant as art work. If a king is shown fighting a lion (as in a mural in Persepolis) it is because it is a legendary event that is commemorated. These are mostly allegorical depictions, i.e., they belong in a known chronicle, and are not outwardly associated to a wall's strength. The portrayals of bulls, horses, sheep and goats in buildings may honor the beneficial gifts of husbandry to ancient societies, primarily nourishment, work and sacrifice. There is some information available about the symbolism of imaginary and hybridized animal forms too[85]. Pegasus and the colossal recumbent sphinx of Giza (2550 BC) are two good examples. The winged white horse (that sprang from the head of Medusa and flew up to the sky) came to symbolize the immortality of the human spirit. The sphinx of Giza is a lion-bodied portrait of the Egyptian King Khafre, in other words, himself in a mysterious spiritual form, one he probably desired to impress on viewers. The location of this sphinx with a pyramid in the background may be that of a sentinel too, however both the idea of an upward flying winged horse in the sky and the sphinx are above everything else, aesthetic wishes of the mind. A nineteenth century painting by John Singer Sargent seems to address the attraction of fantastic design to the psyche (see Fig. 99) by featuring the Giza sphinx as a living structure lifting up his head to look at the approaching chimera. In ancient Egypt there were several theriomorphic forms of important deities such as Bastet and Sekhnet that were hybrid human and feline; Horus, god of the sky, and Anubis, the god that tended the dead had human bodies with falcon and jackal heads respectively. The fact that some animals were regarded as an incarnation or epiphany of a deity in ancient Mesopotamia and Egypt, informs that imagination detected a spiritual life in them, that was connected (though mysteriously so) to the human spirit. The cult animal of Ishtar, the goddess of Mesopotamia, was also the lion, reflecting that ancient people attributed mystic power to it. Some people make contact with animals easily. These animal experiences led the species to domestication and it led Egyptians and many historical people to think that they are connected to spiritual worlds. The cat genus was commonly perceived as 'familiar' which is an animal form of the dead[86], however none of

---

[85] For example, the sirens were originally women birds from mythology that warned of the perils of exploration (in Homer). Anthropologists explain their origin from an oriental soul bird. The centaurs were wild and lawless mythical creatures, slaves of their animal passions (clearly the archetypal alter egos of humans). They were depicted in battles being defeated or punished, for example in a metope of the Parthenon of Athens. Pan was a very common subject of ancient art. His body was bestial in shape having horns, and the legs and ears of a goat. He was antithecal to Apollo's sophistication, and was vigorous and lustful (again, the other side of a man that is not expressed socially). Pan was to be a giver of fertility to herds. The Harpies (snatchers) were probably wind spirits. Their depiction on temple walls in Xanthus in Asia Minor, 500 BC, has led to their interpretation as spirits, as well. The use of sphinxes in temples has been thought to suggest a protective role. The sphinx is well known in ancient near eastern art. From there, the motif migrated to Mesopotamia being endowed with wings. It appeared in Greece in 1600 BC and disappeared in 1200 for 400 years to reappear again. It decorated vases, metal works and temples.

[86] In the representation of animals in art and design, the way people perceive them is important. The themes of the mind that involve animals are linked to beliefs and to our gregarious perception of not just congeneric individuals but also domestic animals. While not all people pay attention to animals, a majority do so. The idea that animals have a spirit is found today in compassionate people. More recently than in Egypt, the radar engineer John Hayes Hammond had his cats embalmed when they died, and he built a grave for seven cats and for himself. His motivations were both his deep affection for his cats and his esoteric research that led him to consider life after death. Similarly, Frederick the Great's last wish was to be interred next to his dogs on the grounds of Sans Souci palace, reflecting his bond of affection for his dogs. We can hear dog owners say that dogs are their 'other half' sometimes, reflecting the emotional identification with animals here mentioned.

this proves that those who thought this way, were animal worshippers, nor that they necessarily had a Totemic animal ancestor. The evidence seems to point instead to a mysterious common manifestation (epiphany) of a deity or familiar spirit in some key animals.

Some typical animals depicted in buildings and other design, were perceived as aspects of deities, and here we can cite the eagle (Zeus' helper) a magnificent bird that inspires respect and came to be the emblem of the legions of Caesar, and eventually of several contemporary armed forces. From these roles we find eagles in buildings, and in the Empire and Federal styles[87]. A lion is hard to tame but it elicits aesthetic interest. Lions and tigers are identified as symbols of leadership in today's linguistic references about some political figures, perhaps because of their relentless and indomitable temperament. Since the unpredictability and danger of a large cat cannot be ignored, lions also bring to mind an important territorial and dominant character wherever they are depicted. The concept of 'king' among animals associated to the lion makes it stand for all the earth's creatures in a cosmological vision. There are several extant coins from Miletus, from the 3rd century BC that show a standard theme, Apollo on the obverse and on the reverse, a lion is depicted looking back at the sun in a *stattant* and *regardant* pose (pausing and head turned back). Apollo became identified with Helios, and *Sol Indiges* in Rome —the sun— from the 5th century BC onward. This iconography of contemplation of the sun by a lion is suggestive of the central role of the sun on all creation when the lion symbolizes all creatures[88]. It secondarily signifies that all animals look up to mankind.

Ancient Rome depicted the Mithraic god *Aion* (an era or age) as a hybrid lion-headed human body surrounded by a snake; three eternally meaningful creatures of creation. Sometimes the lions we see in buildings have human-like expressions. A lion in an ancient grave can oftentimes be an enigmatic evocation of a military leader[89]. We thus understand that some of the lions depicted in tombs, are a leader's animal persona, their valor[90] and perhaps even their claim to royal blood; and they convey to others the legendary character of the soul interred there (see the lion resting next to the chevalier's body in Fig. 78). Perhaps comparably, many tombs of English knights sculpted in stone and brass, show a lion at their feet.

In Italian architecture, gardens and furniture have abundantly portrayed lion masks and heads in cornices and fountains, and these were taken all over Europe and colonial nations took them to other territories. A mask is actually momentous in the idea that the lion form is a façade of a person. Portraits of lion heads and masks have not disappeared from buildings designed in traditional styles today (see Fig. 90). Most of us never notice them, or think of them, but we see them frequently in urban walls, street corners, plazas, fountains, as door knockers,

---

[87] We can additionally cite the cow's sanctity in India, identified with Aditi, Mother of the Gods.

[88] During the centuries before Christianity and in Imperial Rome, the sun was worshipped in Mediterranean regions. Iconography is usually highly philosophical in its connotations. In most of history a lot of input was placed on meaningful imagery that was aimed at reflection. The gaze of the eagle is another representation that conveys iconographical notions. The eagle became used to represent John the Evangelist, and in Christian art it focuses his gaze on the book that reveals the kingdom of God (Schiller).

[89] A Turkish saying teaches that a lion is in the heart of every brave man.

[90] A conqueror was referred to as a lion sometimes. There were helmets of antiquity and the Renaissance, shaped as a lion head, and Alexander the Great is depicted in a sarcophagus wearing a lion skin during a battle. The lion as a symbol of a conqueror is also expressed in the references *Richard Coeur de Lion* and Henry the Lion, both medieval kings. Today a key leader can be associated to a lion, as when Ted Kennedy passed away, there were headings in the news calling him "the lion of the Senate".

and other features, which may be observed when we come in contact with Renaissance and Oriental—styled designs. Lion heads are also depicted in metal hardware, for example, and hold rings for horse harnesses. In the Renaissance Palace of Emperor Charles V a lion with a ring symbolizes royalty and the eagle with a ring indicates the imperial[91] status of the building; but lion heads are also furnished in artifacts for common use that are still sold. They are featured in common furniture brass knobs and door knockers (see Fig. 91). If a lion head or mask has a ring, in the case of hardware, this evidently is versatile to open a cabinet door or drawer. In a knob a cat's head is touched[92], which brings us to the sensory appeal of design, an aesthetic aspect somewhat neglected in research. The sculpture of a lion in hardware, if it is finely made, may recall a cat's head as a pleasing form (this may be true for cat lovers only, however; they are a substantial proportion of people). The designer who chooses a cat head has captured that delight and evokes it in the handles for users. Many figurative forms have been coarsely made throughout much of the modern age though, which reflects that the practice of fine contemporary design abandoned them, and mostly economic crafts have provided them[93]. The human expression of friendliness toward a lion was carved in an ancient Greek frieze in the Choragic Monument to Lysicrates (see Fig. 93) where a lion's paw is held in the hand of a young man and its head is touched[94] (this theme may depict Apollo and the Lion). The attitude of the lion in this depiction is friendly and interactive, and the young man treats the lion as if it was a harmless child. In closing this brief account, cognitive notions lead to a figurative language that is represented in carvings and sculptures, as certainly the mind utilizes evocation to make design an experience that delights the psyche.

But other than the above representations, lions have repetitive and clear placements in architecture and here we can observe a number of unsuspected roles of the cat figure in buildings. It is evident that they have been sculpted in association with buttresses, walls, entrances, passage ways, pedestals, grand stairs and beams (see Figs. 95 to 98). The gate and ground-keeping functions of lion sculptures are well known; but why would their images be embedded in walls, or why would they be portrayed as column bases? A cat's nature is not docile[95], but beliefs endowed inanimate materials with the idea of life and lions were favorite animals, playing key roles as a watchful guard, a palace's or a kingdom's defense and associated ideas. However the thought that a wild cat will protect people is pure fantasy, compelled by the wishes of the mind. The objective of the lion as a support is likely one of vitality and strength and danger to the trespasser. Then, if people —as biped erect beings— were a kind of vertical architectural element, lions —as quadrupeds and prone beings, and alter egos of people— were more commonly connected to horizontal elements. In some vertical applications such as the rampant lions of a palace gate or a coat of arms, lions (sometimes with crown and scepter) are heraldic, often appearing as wardens that hold a shield. Other lions sitting on only their hind legs and holding an instrument are carved in cathedrals and palaces. The holding activity in all these lions is represented in the

---

[91] The motif of a lion with a ring is evident in a throne of AD 518, in a leaf of the Diptych of Magas.

[92] People interact with animals by speaking but communication with them is sensory by touch and behavior intention.

[93] There is very recent interest in reviving and reusing the forms of historical architectural designs in fashion accessories.

[94] There is a saying in Spain that states that God gave man the cat to satisfy his longing to caress the tiger. This feeling is also indirectly suggested in the legend of Saint Jerome (ca. fourth century) who is said to have taken a thorn out of a lion's paw and hence, gained his loyal friendship. There are many paintings of this story, which reflects mankind's desire to tame and befriend a large cat.

[95] Except in the wish to domesticate it.

way that hands grasp something, a human way of doing it. This is not an animal way, as a cat would grab something with his teeth' hence, one more indication that the standing lions, as the masks, betray a human support that is expressed in a feline alternate image.

The sphinx, a lion with human head, speaks very strongly of the ancient fantasy connection between lions and people. Sphinxes were too associated to horizontal ramparts, beams and furniture supports. An architectural element formed as a sphinx can be perceived differently by different artists. An ancient depiction inside a Greek vase shows a sphinx on a short but vertical fluted pedestal (see Fig. 103), which associates a sphinx to a vertical rather than horizontal support, but many sphinxes, as lions, are shown resting on buttresses or lying on the ground under a piece of furniture, or on a long beam. The sphinx as a furniture support can be vertical sometimes too.

Sculptural transformations where a lion foot may fluidly metamorphose into a young woman, or where a column dado may become a satyr or cherub, occur along a vertical support, and perhaps these representations show an iconographical progression that once alluded to sequential life forms, somewhat as in generations, but obeying an ancient fantasy of metamorphosis (or even Totemic descent?) between animal, human and pedestal[96]. This idea of succession is not completely clear though, since a depiction of several images in a vertical support will tend to be stacked, one above the other, in other words, in a succeeding order. But on the other hand, there must be meaning in the depiction of a creature that changes from pedestal to life form.

The lion paw motif is widely repeated for some reason of the psyche. The tendency to mimic design is active here, and also the fact that the mind who does, feels more attracted to repeat this motif, rather than providing a plain post. Something comparable must apply to the buyer of ancient-looking designs. Of all the representations of the lion form, the most significant one is at the same time the most humble, because it is customarily depicted much lower than eye level, and it can go unnoticed: the lion leg of tables and some other furniture. But it must be of great significance because it has not been obliterated by design changes for the length of civilization, having survived from the earliest extant seat of ancient Assyria past even beyond the modernist depreciation of imagery. Ancient society cultivated the lion paw and hind leg, possibly compelled by social pressure to some degree —a connection often made by contemporary writers to account for the use of fantastic design[97]. Social dynamics always affects groups; however, there are other influences that may play a role that we must not overlook because of the always present social pressure. These include the mythological and emotional weight of images on the psyche. Social aspects are related to the purposes of impressing a type of lifestyle on others, so as to gather attention and display importance, while other connotations are related to the rich symbolism of the lion, the fascination it caused in ancient people; and the phenomenon of visual communication through motifs and their mimicry that culture doesn't cease to practice out of a favor for old design. Today, the perpetual lion paw, at the base of a vertical post, may no longer evoke the fanciful lives of ancient social models, but it certainly

[96] The Middle Ages showed a great deal of succession in religious iconography, by means of framed and aligned compartments. The temples of India and china convey the idea of succession of generations, life forms and even many ages in the vertical representation of layered animals and in layered architectural 'growth', which sometimes evokes the constructions of biological organisms.
[97] A Roman sculpture of the Empress Faustina depicts her carriage as drawn by lions; something that is explained by A. Rich as related to the domestication of wild animals for entertainment in Rome.

shows a sculpture of the natural paw of the enigmatic and mysterious lion, that some users of these designs still enjoy, and can certainly associate to classical design. This certainly brings to mind the thought that when ancient people first used lion motifs, they may have done it because they also enjoyed them. Lion paws at the front legs of a chair and the hind legs at the rear, can be observed in some chairs, and this is at least as old as the ceremonial chair of Tutankhamun (1334–1325 BC). See an Egyptian wooden chair in Fig. 117). The hind legs of a lion can be seen in the seat of honor of Dyonisius Eleutherius in Athens at the Theater of Dyonisius, said to derive from ancient Egyptian and Asiatic prototypes[98]. The ancient Roman tripod table from Pompeii, 79 BC also portrays cat legs, though they represent those of a slender cat, perhaps evoking the legs of a lioness or a cheetah. Some four-legged pieces sometimes have their feet oriented in four different directions.

Animal feet became simplified (ritualized as remnant forms) in the cabriolet furniture leg of the Queen Anne period, but the cat paw continued to be constructed all along history, while other furniture legs were introduced and became more fashionable. This was -and is- done for a very simple reason: users acquire furniture with these legs; in other words, these forms have been successful in every generation. They may seem natural, sculptural and artistic; i.e., they evoke a sculptural vision of design as that of the Renaissance, and classicism (see for example the heads ending in a single powerful paw or monopodium in Fig. 114). Some writers have said that they offer a suggestion of status through fancifulness, as they did back in Roman times, and this may be so, but users may commonly say today (spontaneously), that they choose the leg or paw (rather than a plain form) in a part that is known as a 'leg', and is named so[99]. Hence, this is a fitting cognitive circumstance. Indeed, people have also been satisfied with the depiction of a few life forms in general, in both architecture and furniture, as fitting forms. Another conclusion might be that the ancient association of this paw to a building's support strengthened its perpetuity and it became sort of a cultural archetype. The relationship pilaster-paw is very clear in some examples where a corner pilaster ends in a lion paw. There are such pilasters in furniture and in an ancient sarcophagus at *San Lorenzo de Fuori Mura* in Rome, a church that was built by Constantine (4[th C.] AD. See Fig. 118). The presence of these paws, or the ritualized interpretation of this and other animal legs, the cabriolet leg[100], today, points to a collective visual habit of repetition or mimicry of a form that has been satisfactory for very long.

The twentieth century learned to feel insecurity about a taste for imagery in design after its leaders rejected ornament (on the thought that it had represented past aristocracies), and this sense is still around. Up until today, restrained tastes would not purchase a table supported by wild cats, even though the market does occasionally offer 'curious' pieces depicting cheetah sculptures[101] under a glass table and the like. The isolated lion paws however, have been present despite the restrained mentality of the modern age, and they have tended to

---

[98] Lucie-Smith.

[99] In this perception the Renaissance views on taste continue in contemporary buyers.

[100] The cabriolet leg became popular in the west in the 18th century, however, the leg was known in China and Greece. The word cabriolet is French and it means to caper. It was also used as a name for a single-horse carriage, and for the curving backs of the Louis XV *Bergere* chair. It ultimately comes from *capra*, which is a goat in Latin. Because of its name it may refer to a satyr's leg. The cabriolet leg became popular in Queen Anne style as the upward moving society appreciated foreign furniture features. This chair leg is said to have been brought to Europe from China by the Portuguese.

[101] To evoke Africa.

occupy mostly the lowest part of a furniture piece under a table. They sometimes constitute the only crafted parts of 'classic' furniture. This situation is as if this placing of the paw helps craftsmen and users to keep 'irrational' decorative imagery from being too noticeable to critical scrutiny[102] which stands for 'good' taste. Thus this ancient animal motif became somewhat clandestine in our latest century. By this we mean, that there has been no spoken attention to it. The most important point to make about the very common images of culture is that they are always being produced, hence, they are hard to get rid of, despite the fact that they may not be officially in vogue. Centuries have elapsed since the day walls, buttresses and stairs had animal sculptures imbedded in them; five hundred years since the Renaissance masters showed the world how to see deep into creative work, and about two centuries since science dispelled myth. Knowledge and technology have changed dramatically, but the unconscious psyche could not have changed so much in this time[103]. The visual mind must still notice the ancient hybrid, part-pedestal, part-living form, that is still suggested in the lion paw in furniture, even though the verbal mind has not talked about it. The mind must 'know' it in order to produce it and purchase it, even if our linguistic awareness might be in denial of this undying visual habit of the psyche[104].

There are a number of extremely repetitive motifs that have been an inspiration mostly to untrained artisans during the contemporary decades, as many of these depictions are not refined. However, their ubiquitous repetition in all world cultures points to archetypal forms. Among the common images are the moon, the sun, stars, shells, dolphins[105], probably roses, palm trees (a tree of Paradise), the tree of life, butterflies, spirals, lions and other eternally-meaningful images.

As to the irrationality of imagery and ornament, and the identification of purity in design with rational processes (and true aesthetics), objectively speaking, design purity is not free from 'irrationality'. First, purification is not a decidedly rational idea. Second, this perception took us very far from our spontaneous type of perception in a rather short period of time, in its most recent manifestation in the twentieth century and some things cannot be reasoned about it. The theme of purity abounds in beliefs, and all things pure have protected health (as purity in water) to say the least, hence it betrays the advice with survival value. The study of beliefs and rituals by social anthropologists reflects that there are thematic formulas

---

[102] The contemporary accessories market of our perceived rational design age features large cats abundantly in cushions paintings, and statuettes. These come to us as a recollection of untainted wilderness. They are not part of the architectural practice, but they are furnished as accessories because they sell well. Indeed, the wish to love and identify some animals with children (a fantasy of our psyche that is immensely persistent), is present today, though not in weighty architecture or sophisticated design, but in fantasy parks and in the thriving toy and pet design, such as pet salons and spas, dog pools shaped to evoke a paw or other allusive shape, the pet market, pet caskets and cemeteries.

[103] See C. G. Jung for archetypes of the mind.

[104] We must entertain a possibility of an archetype in the identification of a goddess or god with a cat, such as a young lion or lioness (an enigmatic ancient vision of the human form). Not only we have the cat goddess Bastet, daughter of Ra, and the lioness goddess Sekhnet, but also the sphinx, and the mentioned Roman statue of the god Mithra with a lion head, from the 2nd to 3rd century, in the Vatican's *Museo Profano*. In Indian religion, there is a passage where a column is transformed into a lioness. In another chronicle the god Vishnu is challenged to be present in a pillar, therefore Vishnu arises from the pillar as a man lion. The pillar that becomes alive as a cat is a vision that can appear in dreams, even though it may not be a common dream motif. Its expression in both dreams and religion suggests collective elements in it.

[105] An ancient form of Apollo, who sometimes took the form of a dolphin.

in early societies whose intent is to counsel and caution people against evil fates and hence there is an identification of impure systems and the need to establish austerity to prevent them. Their function is to establish boundaries (standards) *to prevent the mixing of realms with evil influences that lead to monsters, hybrids and un-cleanliness that threaten the harmony of the world*. This well-known thematic flavor is somehow present in the vision that rejects imagery. Those who formulated the rejection of imagery did so objectively[106], but the way the value of purity in design was passed down, became shaped by ideals. In our age that substituted beliefs with rational processes, purity as rational reflects the known thematic of 'purity is good'. Whether it is good or not necessarily so, the fact that matters is that imagery is an expression of the creative drive, i.e., an aspect of perception, cognition and its modus operandi. Imagery was once carved in architecture and furniture and it reflects to the scholarly inclined persons the evolution of artistic visions.

---

[106] Moved by function, economy and a new social order.

The Supportive Human Figure

Figures 78 and 79: The ceremonial bearing of a platform or coffin,
a remnant habit of a bearing function.

Left: Tomb of Phillipe Pot al the Louvre Museum shows fifteenth-century
pallbearers. Depiction in pastels of the tomb of Phillipe Pot.
Right: contemporary pallbearers.

Figure 80: Depiction in pastels of a procession with bearers of a platform on their shoulders.

Figures 81 and 82: Women shaped supports: The Erechteum caryatids, Acropolis Athens. Nineteenth century use of caryatids in the Fitzwilliam Museum, Cambridge, UK.

Figures 83 a and b: The ancient association of pillars to the human figure.
The Pillar as a goddess.

Above: Portico of columns with the head of Hathor at the temple of this goddess in
Ta-ynt-netert. *Author's sketch*.

Below: Hathor's face carved on a capital at the temple's site. She is depicted with a cow's ears.
She was worshipped as the wife of Horus at Denderah, and goddess of festivity, dance and love.
Her cult goes back to 2500 BC. *Author's sketch*.

Original portico in a photo from 1930. Columnar human figures supported by lions and ox. Photo of the Museum Oppenheim, Berlin. Detail shows that the weight rests on the backs of lions and bull.

The tall head attire makes the life figures taller so that they become columns. It could suggest that several other ancient head attires of comparable features could have had a similar derivation. Today the portico has been reconstructed by the State Museum of Berlin. Tall Halaf is the city of Goram in the Old Testament. It was important in pre-history and was an Aramean center in the 1st millennium BC. Detail shows that the weight rests on the backs of lions and bull.

Figures 84 a and b: Ancient Syrian caryatids supported by lions and bull at Tall Halaf.

Figure 85: Columns formed as Toltec Warriors.

The columns are at two ends of an area or about 3 square miles in Tula, near Hidalgo, Mexico. Tula is the ancient capital Tollan of the Toltecs', important from AD 900 to 1200.

Figure 86: a and b. Columnar human figures in ancient facades. Photos by Eric Seale

Egyptian male figures at Deir el Bahri, Egypt.

(a)

Twelfth century castle with fifteenth and sixteenth century
Renaissance additions; reconstructed in 1986 after WWII.
Brackets helped by human figures.

(b)   (c)   (d)

Figure 87: a, b, c and d. Top: Columnar giant warriors sculpted at the Georgentor.

George Gate of Dresden's Castle Residenzscholss. Bottom: Female keystone and male
brackets in 18th C. neo-Renaissance building at *Colonnaden* Street, Hamburg.

Transformation in supports

Figure 88: Transformation in Supports.
El Transparente, Toledo
Spain. Narciso Tomé, 1721.

Ritual Transformation and Fantasy. Ancient animals with human faces perform human activities

Figure 89: 2600 BC inlay of harp backside, Ur.

The top frame shows a man who embraces two bulls as if his friends or family (they show the same face as that of the man). The lower stage shows lion and jackal in ritual activities, The two lowest stages show animals playing music. A lower creature as a scorpion is featured with human troso to torso and feet. *Author's sketch*.

Architectural lions not associated with a supportive function.

Figures 90, 91 and 92: Lion heads are still in use.

Left: Contemporary lion mask with pensive expression at condo entrance, Sarasota, Florida. It reminds of a Renaissance face.

Right: Door knocker, Cambridge, UK.

Lion at the Presidential Palace of Warsaw, Poland. The outdoor role of this great lion in the palace is that of a gate keeper. The symbolism of a lion's leadership may become important to the collective psyche in a tragic event. Lion of the Palace of Warsaw covered with flowers on the day President Lech Kaczynski died tragically in April 2010. *Author's sketch.*

Figure 93: A lion's paw is held as if it was a hand, and his head is patted.

Frieze on the Monument of Lysicrates In Athens, 334 BC; von Egle, 1905.

Figure 94: Lions of the processional way of Babylon. State Museum, Berlin.

Lions associated to supports such as walls, columns, buttresses, beams and stairs. The column as a lion in India

Kapitell einer Siegessäule.

Left: Hindi Cave Sanctuary of Mahabalipuram of Madras features an imaginary lion with horns from whose body the pillar grows. See also the temple columns of this chapter's opening image.

Right: Lion associated to a Victory Column of India. Meyers Lexicon.

Figures 95 and 96: Columns blended with an animal.

Palazzo Regia Università in Genua.
B. Bianco. 1623.

Figure 97: Lions of giant dimensions replace the balustrade at the lower steps of grand stairs. Their attitude here is territorial. Palazzo Regia, 17th century. *Source: Meyers Lexicon, 1902.*

Figure 98: Fitzwilliam Museum, lions with juditious (human) expresion top the buttresses of entrance stairs; in a role of gate keepers. 19th century. Museum houses art and antiquities. Cambridge, UK.

Hybrid imagery of architecture

Figure 99: Fantasy of attraction between the architectural sphinx and a human-formed chimera.

The depiction alludes to the human spirit in the winged chimera. It portrays the allure of the architectural sphinx to the psyche; or, in other words, the attraction of this enigmatic design. John Singer Sargent.

Hybrid human-animal forms in walls, buttresses and pedestals

Human headed animals imbedded in supportive stone elements. Ancient Assyria and Persia.

Figure 100: Winged bull of Assyria in vertical slab and gate walls.
Palace of Sargon II; British Museum.

(a)   (b)

Sargon's Palace, Khorsabad. The giant winged bulls have the role of a support in a wall sometimes. Lions depicted as supporting ancient columns on their backs. v. Egle.

Figures 101 a and b: Giant winged bulls with human head imbedded
into the gate walls of Assyria.

Figure 102: Temple of Luxor with entrance of giant sphinxes topping horizontal pedestals.
Source: Baustile et… von Egle, original publication 1905.

Figure 103: Sphinx appears to Oedipus sitting on an Ionic column-pedestal.

In an Attic cup of c. 430–470 BC, in the Vatican Museum. *Author's sketch.*

Figure 104: Herms of ancient Greece. In the State Museum of Berlin.

An early form of a pedestal topped by a human head was the ancient Greek Herm. These were stone land markers that additional to their function, represented the head of Hermes, god of fertility. Herms were placed on roads to protect the traveler and for good luck. The body was a simple slab of square section that tapered toward the bottom to suggest the human figure. The human pedestal idea was adopted by Rome later, and the top was eventually changed for other gods, as Sylvanus or Jupiter Terminus. The form evolved into fanciful torsos of hybrid creatures that represented additional characters. In essence the simplification of the original herm as a sort of tapering pedestal could well be related to the practicality of bulk production, as individuals had the assignment to carry many herms to their destinations according to a text about Alcibiades. However, the idea of a herm blends the inanimate architecture-like form with Hermes' head, which is a life form. The herm idea became an outlet for the mind's fantasies about transformation of architectural pillars into living forms.

The blending of a human head with an animal body is ancient too, as reflected in the Giza sphinx. When the Renaissance school of Raphael (in Rome) researched ancient roman ruins, such as the Golden House of Nero and the thermae, they found hybrid creatures depicted in grottos and they began referring to them as grotesques.

Figure 105: Renaissance pedestal with imaginary hybrid figures at the corners.

Herms in female form with many breasts.

Pedestal underneath sculpture of Perseus with the Head of Medusa, by Benvenuto Cellini, 1554. At the Loggia dei Lanzio. Piazza della Signoria.

Corner bracket imagery of multiple breasts evokes the Asian Magna Mater goddess that became identified in Rome with Diana of Ephesus, 'Mistress of Animals' and goddess of vegetation and the hunt.

Figure 106: Herms support an architectural arch that frames the figure of Erasmus.

Hans Holbein the Younger, Erasmus of Rotterdam. Woodcut, 1530. Vgl, Kat.-Nr. 134.

The herm, an imaginary hybrid, part human part pedestal, became a popular motif of the Renaissance style in Germany, the Netherlands, France and England. The side herms of this drawing are pedestals that become a male torso with a column capital with baskets of plants on top of their heads. They represent the support of the architectural arch which is loaded with fruits. A lion with ring tops the arch similarly to a keystone, and a cherub is above the lion. Erasmus lays his hand on a young male torso (a herm) with the expression of madness or a malicious smile. It is supported by a pedestal adorned at the base with a goat with ring. The frame becomes two sirens underneath, that bend their bodies that metamorphoses into fish tails so as to surround the frame at its corners.

This drawing by Hans Holbein alludes in iconographical way to the efforts of Erasmus to develop a trilingual school for youths in Louvain in 1517, to teach them Latin, Greek, Hebrew, and humanistic studies. Erasmus' work proved to be difficult hence he left Louvain after four years. Woodcut is at the Stadtische Kunstsammlungen, Chemnitz, East Germany.

The engraving's writing translates into: 'The body and the image which Erasmus doesn't see'. The central herm is identified in this depiction with the negative side of men (certainly a herm is a pagan form to Christianity). The name Terminus is a reference to Jupiter Terminus, god of boundaries, the Roman equivalent of the Greek Hermes; in other words, the herm. Terminus also means of course, "the end".

Left: Design for a Mannerist bed.

Right: Fanciful arabesque design shows herms and sphinxes within very slender open walls. The central female figure has a building on her head. Both by Hans Vredeman de Vries (1527–1607).

Figures 107 a and b: Mannerist designs with hybrid architectural, human and animal supports.

Figure 108: The metamorphosis of the corner post from lion paw into fluid human form.

Walnut cupboard showing influence from Jacques-Androuet Du Cerceau and Hugues Sambin; French, 1575–1600, in the Frick Collection, New York. Rectilinear herms appear at the center. The corner hybrid forms have transformed the herm into a fluid hybrid support that contours the bulging of the piece. Author's sketch.

The elongation of the neck in a human-like form has not seemed unfit in art, as in this cabinet or in Hiberno-Saxon interlace and other designs. It is aesthetic as well in women. Long necks in sculpue are observable in ancient Egypt and some African regions where the neck can be elongated by rings.

Figures 109 a, b, c, d, e: Herm pedestals.

Left: Doorway at Wismar's Renaissance Palace,
Center: vernacular herm at Luneburg. Germany.

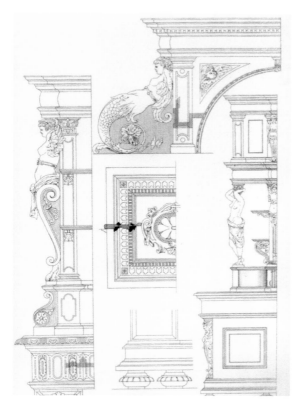

Figure 110: Drawing details for a buffet.

Designs for a cabinet show the herm in
the form of a mermaid.

The mermaid placed at the base of
the cabinet (right) bends to support the
greater weight. A life form (whose fish
tail reveals that it is also a subconscious
form) conveys this supportive
expression explicitly, something that
an abstract building part usually does
not. The representation is one of both
physical and mental support. Design
by Anton Pösenbacher, Munich. In
Schwenke, 1881.

Renaissance revival.

Figures 111 a and b: The expressive satyrs that show how they support the roof of the Zwinger Pavilion on their backs and with their hands, seem to remain faithful to the ancient telamones, Dresden, 1705/8 to 1722. Sculpted by Balthasar Permosser.

Figure 112: Sphinx on a stone beam at the Orangerie of *Marmorpalais*.

The use of sphinxes is related to a new interest in Egypt. The association of a sphinx to a beam is interesting. The palace was built from 1787 to 91 in the New Garden, on the river Havel, at Potsdam. Designed by Carl Philipp von Gontard and Carl Gotthard Langhans the Elder.

Figure 113: Sphinxes as supports of a Russian table.

Twentieth century interpretation of Empire style. Image from *What Antique Furniture*. Used with permission.

Figures 114 a and b: Lion leg with human head as support for a table.

Left: Reproduction of a tripod table design by Francois Honore Jacob. Source: *What Antique Furniture.* Used with permission.

Right: Cast iron leg for a Classical table by Karl Friedrich Schinkel, for a Roman bath at *Charlottenhof* Palace, 19th century. Sans Souci Park. Potsdam.

Figure 115: The hybrid human pedestals of nineteenth century England. Fitzwilliam Museum, Cambridge.

One of the realistic human torsos that support the Fitzwilliam's cupola. The male figure has strongly developed muscles for his task. Comparable Herms are also depicted at Blenheim Palace, England, by Sir John Vanbrough.

Figure 116: Lion paws are the feet of this contemporary Italian cabinet.

Detail of corner supports also displays a devil's head.
1970s Renaissance styled cabinet; privately owned.

Figure 117: Reproduction of Egyptian woodwork chair, New Kingdom. All lion paws oriented in the same direction. Image source: *What Antique Furniture*. Used with permission.

Figure 118: Pilaster becomes lion paw.

One of the pilasters of a sarcophagus at the church of San Lorenzo de Fuori Mura, Rome, ordered by emperor Constantine in the 4th century AD.

Chapter 7

The Intuitive Concave Forms

The search for large spans was a prerequisite step to create the soaring vaults of history. In the continuum of discovery and transformation observed in ancient construction, the true arch had first permitted large interior spans in stone. Vaults were in existence before Rome adopted them, but it was Rome that shaped vaults into very significant concave designs, in particular by the utilization of the groin vault and the dome.

While a vault is ultimately a means to create larger spans, the mind has many pursuits that attempt to adapt design and a vault is also a design instrument that attains concave enclosures. Ancient Rome molded bowl-shaped spaces by manipulating the arcuated method and designed spheres, apses, exedras[107] and vaults. Its production displays abundantly surfaces that curve in plan, and vertically, in half-spheres and quarters of a sphere as the Pantheon, the *thermae* and the basilicas (see Figs. 43 to 45). We can read in history that Rome's concave forms are based on the culture's admiration for the circle and the sphere[108]. The Roman culture created more than just circular designs as it left us cavernous architecture, in particular in its baths. The concave designs of Rome also include smaller structures and chairs and the classicism of Rome is foremost one of plastic appearance. Renaissance and neoclassical buildings adopted Roman classicism, with some emphasis on the vaulted basilica, but a truly cavernous appearance reached a zenith in the Baroque style of Italy, Southern Germany, Austria, Prague and Spain, while it was not uniform in neoclassicism, and this may have been an offshoot of the appeal of a rationalized (controlled and calculated) classicism of the age of reason. The Italian Baroque churches of Bernini explored the syntax of the Pantheon in Rome, where the space of the dome dominates the total design. Bernini's domes are oval interiors which are made organic through sculptural treatments. The Middle Ages had shifted away from classical proportions, tending toward verticality and pointed syntaxes, but also produced tall and cavernous vaulted spaces, including vaults in miniatures that surround the human figure (see the figures of chapters 9 and

---

107 Larger niches sometimes placed toward exterior gardens.
108 Concave forms, like circles and arches are fitting to the mind.

10 of this book). Rome was the first to produce an enveloping stylistic syntax (encompassing many crafts) in its classicism.

The use of stone in small units of masonry can be observed to be one more detail in a tendency toward concave design. Masonry resists compression well and can be stacked up in tall heights; and in the large ancient Roman basilicas and baths, the size of masonry is sufficiently small —a finer 'grain' compared to post-and-lintel elements. Though stone masonry is equally unbendable as stone beams, its small size makes it more pliable to be molded.

Circularity pre-existed in Greece, in the plan of temples, but it changed the design scene completely in Rome and its colonies. Architecture began its progression toward a sheltering curved form that had continuity in Byzantium, Ravenna and Armenia. The enveloping niche with a human sculpture speaks of a personal architectural space for a human form (represented by a sculpture), in its scale, depth and its concave wall, rather than a rectangular choice of planes. In looking at some artifacts of Rome one can't help but notice the 'tub' chair (see Fig. 121) though this design does not seem to be an offshoot of cylindrical walls that are miniaturized in a chair, and any such influence would be secondary, nor does this chair display any structural décor, but it has a concave back for the convex human back, possibly as a trial design toward the enhancement of comfort[109], even if our contemporary point of view does not see comfort in it. But because a chair is in direct contact with the body, a contouring form is intended and visualized by a designer (the circumference of the body of a sculpture may also influence a niche too). The idea of a surrounding and curving chair back can be said to have had had at least an ancient precursor, the seats of honor in the Theater of Dionysus (330 BC) which were also cylindrical. This throne seat is said to be related to Asiatic models[110]. Then, the idea of fitness of a chair for the body is observable in the *Klismos* chair of Greece (see Fig. 122) which has a narrow splat that curves on the back. But at any rate, all concave, surrounding and cavernous architectural spaces, chairs and also some carriages, may speak of designs that progress toward a form that envelopes the body.

The mind's transforming and adaptive pursuits can be seen in action in the crafts that depict structural elements as decorative motifs. These crafts tend to simplify structural elements and often, only key elements are shown, and even their proportions may appear altered, looking disproportionally small, or overly elongated, or be of a non-constructible slenderness or openness, giving an impression that the artists didn't understand the proper relationships of mechanics and materials, or even perspective or the needs of utilitarian architecture. Distortion is common in medieval representation and it can appear to some observers to have resulted from crude skills. While medieval skills at proportioning sculpture are certainly less skilled than the Renaissance perspective, the medieval distortion gives place to the manifestation of a remarkable representation that matches the human configuration and is reached by shaping an arch further, in the foliated direction. Being the Middle Ages a period of strong faith, this happened through the channel of religious symbolism. This artistic outcome seems to connect a deep (or sightless) sense of the mind involving the human form, with arches. This significant sheltering notion was achieved in the crafts.

---

[109] It could have been -as we say today— an ergonomic vision of the back panel that surrounds the body's back. There is no distinction between comfort and some sheltering effects, sometimes, and this chair seems like an answer to both. The term ergonomics was coined in the late twentieth century to refer to reduction of worker fatigue through design.

[110] Several authors, for example, Lucie-Smith1979.

Some designed items (which do not have to be considered aesthetic by a contemporary viewer), may be perceived with interest and even empathy because of their connotative qualities, such as the above-described human-like outline in the structure of an arch. The innocence of the Middle Ages helps us see a vision of minds that were untouched by sophisticated training so to speak, however they were highly talented. Perhaps today, a similar effect may appear in a few vernacular crafts. But coming back to one of our important topics, this process takes place largely through the migration of the structural arches as a religious motif.

The simplification of architectural elements in sculptural work may reflect those aspects that were recalled by a craftsman from a building, when carving a miniature, which he did not have in front of him— as the sculpted miniature, may be partly imagined and partly recalled from a structure at a different location. This effect was involved in creativity in certain past epochs freer from schooling than in self-conscious periods. This simplification of details seems to have been the case too, in buildings depicted in manuscript illumination. The ritualized depiction of structures in medieval miniatures (both in religious texts and in sculpture) shows extraction from imagination and memory, and the transforming action of the mind on the proportions and the presentation of structures (see the art shown in chapter 10 of this book). In other words, in all likelihood the simplification and also, some adaptations that structural motifs show, are a ritualization of a building's forms into an imaginary structure. It shows us the building's most remarkable characteristics in part transformed according to an artist's recollection, or those elements remembered[111], where the scale of architecture is miniaturized. The ritualized elements nevertheless are readily identified as representations of structural elements (arches, roofing, columns, etc). There is additionally in them the important human-formed architectural frame. Why concave forms, or, why does this vision of architecture tend symbolically to the human body? This phenomenon has the flavor of a key event in the development of human shelter, that we call architecture. What was architecture to early people? It seems to represent a vision of protection, that was (or is) in the deeper mind and perhaps it comes out as a reply to the long human search for architectural forms.

An Egyptian sarcophagus has the above type of design too, and its body form is an intended design to fit a body inside, as it is a personalized container of limited dimensions whose human-shaped exterior and surface reveals whose remains are inside. A sarcophagus must be also related to the desire to commemorate the body in funerary sculpture and certainly to preserve it. The Egyptian sarcophagus, the tub chair or an ancient *currus*, a Roman carriage that has a surrounding shape[112] are influenced by a streamlining intent after the human form. Some of these designs had to be both light and fit for the body, such as the carriage that must reach speed. The progression toward an enveloping form for the body in a large-scale architectural space —that has no contact with a body— is certainly not the same, in the fact that it is most commonly not thought about. In this latter case, the concave outcome seems to be an intuitive artistic development.

---

[111] Lewis Kausel, 1982.
[112] Ancient Rome created the *currus* for two persons.

The Design of Cavernous Interiors, Distortion and the Idea of Movement in the Baroque Age

Abbey of Wieblingen, Christian Wiedeman, 1744. The ceiling paintings use *trompe l'oeil* effects.

Figure 119: Attempt to design movement and architectural elements that flow into one another.

Viertzenheiligen Church. Balthasar Neumann. Near Bamberg,

(a)

Viertzenheiligen Church

(b)

This soft whitish surfacing with organic decoration is observed in some porcelain of the same period. The Rococo design blends walls and ceilings as a continuous organic surface.

Began in 1740s along with the abbey church at Neresheim in Swabia. The church Vierzehnheiligen (the fourteen holy helpers) of 1743–1772, was to have, as its central element, an altar built over the spot where the 14 saints known as the Helpers in Need had appeared in a miraculous vision. At first it was thought that the church should be built on a central plan, but Neumann's design for a longitudinal-plan church with the altar under the dome over the crossing was accepted. The builder entrusted with the construction began the chancel incorrectly, and Neumann had to step in and alter his plan, so that the altar was now in what would have been the nave. He skillfully resolved this unfortunate situation by breaking the nave up into ovals; in the center of the largest oval was the altar, thus giving the impression that it is, indeed, in the center of the whole edifice.

1) Reuther, H.: Balthasar Neumann - Der mainfränkische Barockmeister, Süddeutscher Verlag München, 1983.
2) Gale Encyclopedia of Biography: Balthasar Neumann, Church Architecture

Interior plasterwork detail of the Basilica St. Emmeram, Regensburg, Balthasar Neumann.

This concern with blending a central-plan church with a longitudinal nave, so fortuitously worked out at Vierzehnheiligen, found its fullest expression at *Neresheim*, Neumann's last great church (1747–1792).

(c)

Figures 120 a, b and c: Space of a nature-like quality achieved by shaping and decoration.

Figures 123 and 124: Arcades in iconography (manuscripts) can be
extremely slender and open in typology.

The tiny vaults seem to represent the inner surface of a mind in a dream. The blue color identifies these vaults and roofs with the sky (heaven). Generations of the past are depicted in an attic's elements and on the roof, while scenes from Mary's life are found within walls and spaces.

Left: *Speculum virginum* Manuscript, Jesse and House of Wisdom; late 12th C., British Museum. In Christian iconography the seven pillars stand for the Temple of Solomon. The central pillar is genealogical showing Jesse at the bottom, the branches of his tree, Mary at the center, and Christ above. The heart shaped leaves are a representation of the seven gifts of the Holy Spirit. The right and left buildings in the iconography of Christ's genealogy are usually synagogue and ecclesia. These buildings are located high above, supported by slender columns and arches.

Right: Virgin and child. Manuscript written for Isabel Stuart, daughter of James I of Scotland, 1445-50. Fitzwilliam Museum, Cambridge. The slenderness of the supports in this depiction would be impossible to achieve in masonry, the medieval material of choice. However, there was a clear tendency in medieval days toward slender architectural elements that sometimes resembled tendons and even innervated structures. Their color may give us some other information as to the kind of anatomical structures that it may represent (should this be the case).

The Madonna occupies the height of the interior of this castle-like church. Hence, somehow this is seen as Mary's structure or her personal shelter. The blue vaults and Mary's blue mantle are identified in medieval iconography as representing Heaven. The scenes are from her life and her marriage. Her ancestors or the prophets who lived long before she lived, and prophesied the events of her son's life, are here depicted at the roof level.

As in many dreams about architecture, the building depicted in this illumination is mostly white, and its walls are actually, compartments (spatial relationships) where scenes of other times and persons related to the central individual can be seen.

It is interesting to study the distortions of space and supports. They reveal imaginary features for an ideal architecture that in this period had heavenly characteristics.

Chapter 8

Cognitive Culture and the Ornamental Application of Structures

The structural motifs that history's design used in objects left behind the constraints of mechanics and sometimes additionally, their architectural proportions. They additionally left behind their original materials and became reinterpreted in a different medium such as wood. Reused or reapplied structural forms may advance beyond the simple environmental mimesis of design, and be beautified beyond the refinement level of the inspiring model. If an object displays an attractive appearance, it has high chances of being reapplied in more artifacts thereafter. The aesthetics of structure has been inspiring and it could be enhanced in its ensuing application. The beautification of a structural image comprises stylization of the essential form, or its treatment, or both. Perhaps aesthetic enhancement honors (or exalts) an original system that worked well. An explanatory example from history follows in the next paragraphs. We believe that this and the other examples described in this book can help clarify some of the purposes of stylization and ornamentation in design.

We cannot think of a better example of reapplication of a very utilitarian structure as an outstanding indelible image of all times, than the reuse of the arches of a slender and splendid stone aqueduct[113] as interior arcades, in the Mosque of Córdoba (see Figs. 125 to 127). The two structures, aqueduct and mosque, are separated by some ten centuries, and are also the work of different cultures (ancient Roman and Islamic), however, the permanency of an aqueduct in the landscape, and the occupation of Spain by Islam certainly permitted a direct inspiration, as an aqueduct continues to offer a potential model that can inspire later design as long as future generations can still see it. In the interior aqueduct paradigm, the aqueduct's multiple arcades have been interpreted in brick masonry and plaster (or *escayola*), that was refined and enhanced by texture, foliation and by coloring the material. This exalts the arcades by beautification. An aqueduct was meant to give a

---

[113] Probably the *Los Milagros* aqueduct of Mérida.

service, thus it was strictly utilitarian both in origin and objective, but the ancient Roman builders usually produced a harmonious composition in structures, even if these were destined to provide only a service as a public utility (aqueduct, bridge or cistern).

The aqueduct gives an impression of an uninterrupted, well-balanced rhythm of arcades, where this repeated motif or arcuated pattern arises from the need to make a structure very long. The idea of a sequence of superposed arches as in the *Los Milagros* structure (where function is inseparable from structure) was later applied to the Mosque to support the ceiling. In reapplying it, the *venustas* of the arcade was enhanced further by transforming its appearance. The natural *venustas* of an arched structure informs something symbolic that is inherent in its roundedness. This application thus transformed the semi-circular arch into a horse-shoe form[114] and into a lobed (or scalloped) version[115] of the arch. The lobed form reveals attention to design from molds, and perhaps from this type of form, the convergence of an evolved scalloped arch has made its appearance. The arches with curving segments that surround a circular space, evoke the contour of the human head and body outline, and the lobed smaller segments can evoke (in a fully evolved lobed arch) a mold of the brain[116] (see horse shoe arches with human figures in their interior in Fig. 129, and an arch imagined and drawn by the author after the outline of the brain in Fig. 130, that is comparable to the lobed or scalloped arch). The evocation of the brain's lobes does not seem to have been arrived at in awareness of this design[117], in particular in Islamic design that rejects the human form. A case of a directly inspired design is very unlikely.

Once the form of a structure is ornamented and enhanced, it may continue in culture along a path of further decorative application in interior crafts. Interior reapplication of structures lends itself to a more exquisite treatment than when these forms are exposed to the outdoors where such refinement would be damaged by the ravages of climate. But, just as in the Pantheon, an Islamic interior is also soft and uses fine materials whereas the exterior can be quite plain.

Despite the fact that refinements were desirable to history's designers, the Spartan design ideals of recent times have not been enthusiastic about decorative treatments. The contrasts that arise in the aesthetics of different periods are an interesting topic, in particular here, under the light of the fact that history's architects did not disagree with the practice of embellishing buildings for most of human history. However, twentieth century architects were trained to be disciplined so as not fall for the 'facile' approach of decoration, where an idea that design must rise above popular embellishment is also implied. In the modern axiom the structural design of an aqueduct of a kind is the best design, and it is indeed a meritorious static design and a model form for later arcaded structures. Structure is the essential design phase for any building to

---

[114] The horseshoe arch can be traced in extant early Syrian and European examples.

[115] The foliated or lobed outline has been traced back to gored domes through Mesopotamia and to Byzantine, and ultimately Roman, shell niches, Hersey, 1937. There are important analogies and linguistic roots that indicate cognitive parallelism between a scallop shell and a skull as well, see Lewis Kausel 2007. Lobed designs occur at the level of floor plans as well, as that of *Minerva Medica* and the Roman baths of *Conimbriga* in Portugal.

[116] Lewis Kausel 2007.

[117] The intended evocation in the lobed design is probably a representation of a scalloped shell. The scallop is sculpted in niches, and it inspired a type of structure, the gored dome. Shells were depicted in many monuments of Islam. They also have a high variety of symbolic meanings around the world, having been used universally by human and pre-human groups. There are meanings attributed to shells that pertain to funerary customs, namely, protection, origins and rebirth, and this symbolism may also be related to their representation in architecture. For the many cultural symbols of shells see for example Eliade, 1969.

exist and last. However, does this indicate that other design phases are not good or necessary? They are certainly not necessary for a building to stand[118]. However, a transformed version of an arcade speaks of something else that is also significant as we have just seen. The embellishing exercise strengthens the cognitive power of design[119]. It also refines the structural outcome, and perhaps through elaboration it can reach what the design in question can look like under the lens of the inner psyche of the period. The mind's expectations measure aesthetic qualities and find the looks that are delightful. The interior application of the arcade, in the horseshoe or in foliated form, is a step further in an interesting direction of imagination that sometimes mirrored its own image in the design of history. Though this phenomenon might be seen as a case of unintended anatomical mimesis that the mind creates, the aesthetics, the genesis and possibly the purpose of this design event are yet to be better understood. The mind arrives at design solutions that can parallel natural solutions even without intending to imitate nature. Is the genesis of this design tied to this purported ability of the mind? Parallelism can sometimes be observed in other contexts of knowledge.

An ancient practice of miniaturization of architectural forms acquired momentum in the Middle Ages. It was common in the Near East, North Africa, Europe, India and the Far East, and possibly in other corners of the world that were not at that time influenced by any of the old world regions. Miniaturization is somehow inherent in the reapplication of the images of structures (or form migration), hence, it is an idea that can arise in the need to adapt and shrink designs that are deemed worthy of being repeated as designs for other entities[120]. Thus, in the fluid transformations of imagination, the image of a building can be shrunk to miniaturized scale and it can be distorted and reshaped. A structural form that is miniaturized frequently comes closer to human scale than a building's natural size, and sometimes it is even smaller. In the example of the reuse of the Islamic *muqarna* or *mocarabe* vaults, as tiny decoration, as in wall alcoves (*hornacinas*), the migration of the vault decoration to walls is an event related by trade, and it developed at the same site, and certainly within the same time period as the ceiling vaults. These miniatures clearly reproduce or portray the aesthetics of the *mocarabe* vault as a decorative pattern for surfaces. They are a case that fits the idea of mimicry in a miniature. The aesthetics of an arched structure stimulates other minds to the point of making them wish to represent it[121]. This seems to be the case too in European medieval art where many craftsmen provided visions of the vault in decorative versions. There is oriental influence in the sculptures of medieval European art however, miniaturization itself is probably a phenomenon arising in the wish to represent. This wish to represent can be seemingly stimulated by the aesthetics of the functional vault.

This practice in Islamic interiors multiplies miniature vaults and sometimes offers a membrane-like semblance in *escayola* decoration. Judging from some rare depictions of religious persons in Islam's art, the halo of saints has a flame-like configuration that some of these tiny vaults seem to also reflect as a complementary outline of that halo (complementary as the form of the Gothic

---

[118] A mutually exclusive logic is however part of our tendency to understand visual culture in highly simplified ways that are not always faithful to its reality.

[119] The designers of the tenth century, for example, were still relying heavily in their intuition, so that this fact might account for a way to arrive at such parallels. Our generation tends to neglect this exercise, since we count with so many other resources.

[120] And as we have seen, miniaturization also arises in the idea of mimicking a religious building in a small container.

[121] Mimesis and aesthetics are many times connected in the mind.

vault and a bishop's miter). The human figure seems to be important in the subliminal domain of architecture, according to the observations of this author, and it may be manifested subliminally, even in a faith that rejects its representation, which is indicative of an unrealized process of the inner psyche. The brain configuration in structural design (it is also manifested in the Pantheon's interior and the vaults depicted in the artwork of the author that reflect the brain) is a very interesting case of design. A structure (arch or vault) is the basic catalyst of these representations.

A further glance at the smaller crafts of history bearing structural imagery reflects an alternate application of some building-like forms by culture that was highly significant: namely to represent concepts needed to indicate a honorable place for revered individuals, such as a king or religious person. For example, the concepts of honor and veneration have been indicated in the ancient idea of enthronement. Many historical thrones have features that are clearly architectural[122] however, they are certainly not utilitarian but they obey an iconography of imaginary architecture (see Figs. 135 and 136). Religion and architecture present a case of mutual influence. Gates, cities, towers, fortresses, shelter, temple, synagogue and ecclesia, are widely used as representative pictorial ideas of sacred cities and divine refuge[123] (see Figs. 1. 123, 124, 135, 137, 142, 152, and 154 to 156). In Fig. 1 ivory holy men have a personal space under the eaves of an umbrella dome (the dome being heavenly). Illuminators of manuscripts gave these architectural forms to the religious concepts that were mentioned in the Scriptures as gates, fortresses, etc. In this type of application, we should begin to see that society resorts not only to words, but also to its material culture, to express meaningful concepts. The faith concepts of majesty, paradise, heaven and glory, are indicated by canopies and compartments bearing structural elements in these manuscripts. These are spiritual visions that the mind imagines and designs. It is known that illuminators of manuscripts interpreted faithfully the older religious texts and gave a pictorial depiction to heaven, paradise, etc. However, spatial and sheltering concepts are aspects of a sensory logic that we carry about space and its relationships. These are cognitive concepts that were given architectural form early in buildings, as mankind began to subdivide and shape its interiors.

It is important to acknowledge the architecture-to-analogy exchange of religion, since it reflects that images do not just migrate from architecture to object (i.e., from design to design). In all likelihood, concepts inherent in cognition feed architectural forms first[124], and once they are developed in a material form, they in turn offer their forms for religious metaphors. These metaphors are used in written descriptions and are also portrayed in illumination; and from here they pass back to the sculptural crafts of buildings. Then, the forms of buildings illustrated in early religious texts were carved or painted in altars "to the letter" of the sacred text, thus these illustrations passed back to building in a process that carefully repeated them for religious reasons. In the case of architecture and religion, there is a centuries-long mutual feedback

---

[122] Elaborate chairs or thrones indicate a place of honor and were many times built with canopies, and were placed higher in a room, so that the user of a throne had to climb some steps to reach it. As background information of these features in a throne, the bible describes the throne of Solomon, and it mentions steps that led to it.

[123] Some place-concepts of faith that have no physical counterpart, such as Paradise or Heaven, belong in the collective imagination of people. They are created by the mind and give human hope for relief from suffering.

[124] For example, the idea of doorways or passage from one enclosed space to another type of space, and the ideas of higher and lower levels, stairs, and a cover or shelter above a space, must all be first cognitive aspects of the mind that were given architectural form.

between culture and architecture that increased in complexity because of the use of iconography as a conceptual language. An added significant dimension of the transaction faith-architecture is that architectural form becomes associated to numinous and mysterious concepts and events. This should illustrate how the image of architecture is endowed by people with connotative value, and perhaps a good term for this content even outside of religion is still the 'iconographic value' of architecture, as it is a visual form that can stand for important concepts of society. Structures in particular have a conceptual life in much common secular metaphoric phrases in implying ideas of support, connection and shelter. Examples are ideas such as the "pillars of civilization", the "foundations of knowledge", a "bridge to the future", or "a roof for the family" as necessary provisions of dignity. Structures are associated to strength, reliability and the basis of ideas, and in the case of bridges, to junctions, defining moments as between a past status quo and something that will change. Hence, many structures become leading and meaningful concepts. There may not be in the mind a particular pillar, foundation, bridge or cover, but a very general image that can enfold all pillars, all foundations, etc. Society therefore identifies its architectural capital; it spontaneously generates an iconography associated to some types and makes good use of it ending up cultivating its building images.

The concepts of imagination have a tendency to migrate and become transformed in culture. Several practical trades have conveyed something aesthetic to their creative output. This transformation has been focused until now on form and ornamental manipulation. But columns, arches, pediments and other structural forms are also subject to transformation, even though such elements were not in origin intended to be connotative or be used as embellishment. In fact their principles are sometimes seen as transgressed by the crafts. The design habits and preferences of users can become significant when everybody tends to proceed in one direction, as culture derives cognitive syntaxes from structures and transforms them into stylistic concepts. Hence, imagination consciously or unconsciously may break the understandings of the schools of thought in design that attempt to establish the domains of design. It is not only the craftsperson that breaks the rules, it is also the user who reads aesthetics in the decorative use of structures.

Figure 125: The ancient Roman aqueduct Los *Milagros* of Merida, Spain.

Figure 126: The arcades shown in the horse-shoe shape of Spain.

Cordoba Mosque arches of the 10th C. This intereior is thought to have been inspired by the multiple arcades of *Los Milagros*.

Figure 127: The arcades become lobed (foliated)
arches in the Córdoba Mosque, 10th C.

Started under the Umayyad rule, at the end of the 8th C., it was built on top of a Visigothic cathedral. The mosque was mostly built under the Berber Almohad rule and its style is *Mozarab*. Some sources state that masons used wooden tie bars to support the arcade but they didn't work, so they replaced the bars with brick arches that were derived from the *Los Milagros* aqueduct of Merida. The Almohad Great mosque of Tlemcén's in Algiers, 11th C., has foliated arches that were inspired by those of Cordoba.

These arches have the contour of a human form. Wall decoration represents an outline of a human figure with an aureole at the head, as in a row of saints.

Figure 128: Single row of horse-shoe arches in wall of the Cordoba Mosque.

Figure 129: The Horse-shoe arch as a compartment for the human form.

Marble bas relief of *San Genis des Fontaines*, Roussillon, France, eleventh century. Illustration by author, 1982. Source: *Design* and Nature, 2007. The form of this arch had been used in Syria, Spain and France since the early centuries AD. In the above relief the architectural imagery tightly frames the bodies of the human figures represented (six apostles). All the capitals of the columns are located at the level of the neck and the lower part of the columns taper and bulge according to the shapes of the bodies they frame, while the three-quarter circle arch echoes the head. This representation clearly shows the arch and columns as a compartment for the body, with complementary outline, and the relationships of supportive and supported elements, both in the human body and in the arch. This transformation generates an almost matching space for the human figure, like a mold or a jigsaw puzzle, and the columns appear as if made of a bone shaft.

A possibility that a carver intended a form that echoes the human outline in the arch, seems far-fetched. Similarly, any awareness of biological parallels in creating these arcuated systems as a compartment for the human form is too remote, hence it is possibly intuitive. Thus, the reason for the appearance of the human form in such close association to structure (arch with columns and capitals) must reside in a subconscious and intuitive logic of protection. This is a fitting niche for the human form, as if the constructions we have been calling 'architecture' (a technique of a higher order) has a subconscious analogy to an idea of a natural-looking structure for the human body, like a shell to the clam it houses[125].

---

[125] Lewis Kausel, 2007.

Figure 130: Design and illustration by author after a cross-section of the brain. Shows remarkable resemblance to the horse-shoe scalloped (lobed) arch. It also evokes a shell. *Author's Original art work. Design and Nature, V. 3, 2007.*

Figure 131: The Friday Mosque vaults. Morocco. Illustration by author shows parallelism between a Moroccan type of mocarabe and the brain's sulci. *Source: Design and Nature, volume 3, 2007.*

Figure 132: *Sala de Los Abencerrajes*, Alhambra, Granada.
Membraneous appearance.

Profusion of ornamental *mocarabe* vaults hang from the eight arms cupola. The current name of this room comes from the sixteenth century. Little niches are bracketed out one above the other. Miniature squinch arches connect rows and tiers of these tiny niche shapes forms.

This decoration developed around the beginning of the 12th century and reached exceptional developments in the 14th and 15th centuries. The richest examples are observed in the Moorish work of Spain as the example shown above.

Figure 133: Miniature decorative lobed arches of Toledos's *El Transito* Synagogue.

Compare this wall decoration to the lobed arches. The material is escayola, a kind of plaster

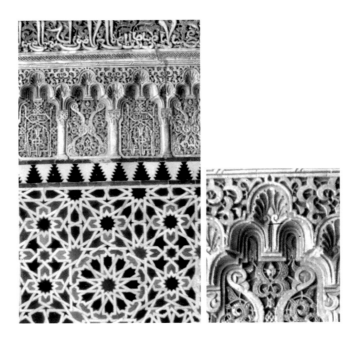

Figures 134 a and b: Escayola treatment for walls, shows miniature *muqarna* vaults above the tiled wainscots, Alhambra, 14th C.

Compare this miniatures to the vault type of Fig. 172.

Honorific Seats Expressed Through Ritualized Architectural Forms

Figure 135: Madonna Enthroned, Tempera on Wood, Cimabue, about 1280–1290, Galeria Uffizi, 'Giotto and the 13th Century', Florence.

Building-like honorific seating piece with arches and a high open typology, depicts Mary's majesty.

Figure 136: Madonna Enthroned.

Enveloping building typology as that of a stadium, like the Colosseum, is suitable as Mary's throne.

Figure 137: Arched forms and arcades evoke a protective
concept of a monastery as 'above' in heaven.

Ritualized architecture above the figures of holy persons show a sheltering foliated arch.
Reliquary of Saint Aemilianus, San Millán de la Cogolla Cloister, Rioja, Spain. Author's drawing
from 1982.

This motif is related by theme to other medieval representations that show arched architecture
surrounding a human group. This carved depiction is found in a reliquary of Spain. The top part
of this theme recalls the roofing tiles and arcades of the cloister of *San Millán de la Cogolla*, where
the monks represented lived. It represents the framing of the monastery above them. The tallest
person, Saint Aemilianus is framed by the central and tallest arch.

The persons depicted are St. Emilianus with St. Aseolo, St. Atrocio and St. Sofromo. The
reliquary dates from the times of the first Benedictine cloister. It was founded in AD 537 by a hermit
who died in this place. Today's cloister dates from 1504–40. It was a common medieval custom to
represent important sacred personages larger than the rest of the individuals depicted. Thus St.
Emilianus is teacher and pastor and the central religious person of the three other saints, in this
iconography. This cloister was an important center of crafts in Spain.

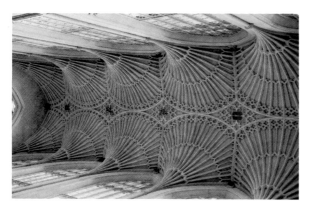

Chapter 9

The Mind's Authority in Design

The construction trades of history were substantially challenged by the pedestrian means and procurement of materials of history that impinged on how objectives were met. But despite these circumstances, inadequate resources did not prevent builders from creating objective forms that still are unsurpassed today. Masons used ingenuity to handle methods and materials so as to reach aesthetic outcomes through them. Sometimes the equilibrium of slender structures seems almost unachievable for the know-how of antiquity and yet a mastery of proportions was then possible. Many times it took more than a century to finish some constructions using superhuman efforts, but they were achieved, and nothing seems stronger today in the attachment of culture to these buildings and their preservation, than their superior quality, their stunning designs and craftsmanship.

The control of slender heights so that a structure doesn't collapse from storms, as in a slender aqueduct, denotes an engineering mind, capable of calculating and fine-tuning the structure's features. The fact that a slender aqueduct like the one in Segovia, of more than 10 miles of length has stood up with integrity among the structures of this type (some minor reconstruction was done in the fifteenth century) speaks of its nearly optimal proportions for the material used, for its mass, earthquakes, warfare and the settlement issues that a long aqueduct faces. These static forces in such an exposed large-scale structure rule the form achieved. Ultimately the way in which a vertical form behaves in wind must be in part sensed personally by builders acquainted with the steering of ships and athletic activities that topple the body. These sensory dexterities and the sense of equilibrium have been experiential cognitive resources of the mind that have helped guide some of the formal features of a structure in equilibrium.

We must not fail to remark that as the function of masonry arches became intimately understood, height and slenderness became enhanced until they approached limits in some daring medieval constructions, such as in the 158 feet tall Beauvais Cathedral, over the choir (see Fig 19 b). The arcades of a tall aqueduct are somehow in the background of the medieval nave arcades, which are also arranged in a sequence. Vaults had reached a long history in the twelfth century, show a subliminal composition of the brain in these ribs as in the Gloucester cloisters and other English cathedrals (see Fig 49).

Medieval architecture offered a spectacular vault that seems to have become an artistic obsession among craftsmen and clergy who ordered their repetition in aediculae. This vault reached a level of profusion of small arches never before seen in European architecture. The migration of structure to decorative treatments, interior woodwork, art work and detailing was plentiful in the Gothic age itself and beyond, as Gothic arcuation it is still practiced today, primarily in religious design that includes doorways, windows and pews. In the Gothic cathedral many saints are columnar in their location at portals (see Fig. 152), in their exaggerated length and their rigid positions. The anthropo-mimetic pillar concept is thus also suggested in this age, in the fitting of a sculpture in a long supportive element.

Figure 138: A celestial vision of the structure of the cathedral of
Notre Dame de Laon, 12th century, France.

The Gothic cathedral of Laon is not devoid of a semblance. It is configured as a Latin cross as most churches. This is a figurative rather than a supportive requirement. However, the main aesthetic feature of the interior of the cathedral of Laon, is not the pure cross layout exactly, but the visual effect of its sequential and slender ribbed vaults in the layout. At Laon, we notice that the ribbed structure follows a nearly perfect order and control.

The vaults of medieval manuscripts are colored blue and the mantle of Mary symbolizes a celestial shelter that is sometimes depicted with stars. In this digital art work the stone of Laon is colored blue. Based on a photo entitled *Picardie_Laon4_tango7174*, a Wikipedia Commons image.

Figure 139: Star Vaulting. Ely Cathedral Lantern.

A Norman work tending to a celestial concept; the lantern was made of timber, by William Hurley in 1334. The ribs become tendons as those of the hand and arm.

Designing is giving to materials the features of a concept, whether this concept is a mechanical apparatus or a natural form as a star. The mind creates by first generating a concept and then by translating it into materials. The intellect produces an outcome as close to what is imagined as can be achieved with the means at its disposition. In looking at this star of tendons, the medieval builders did not separate its structure from its concept, and in the Middle Ages, there are no good reasons why they should have.

(a)

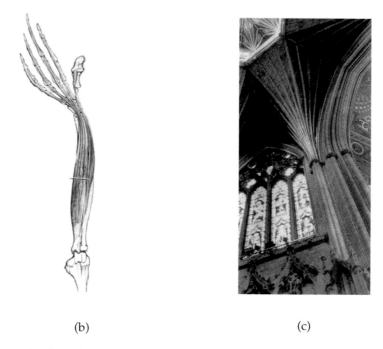

(b)   (c)

Figures 140 a, b, and c: Star Vault and the hand's tendons. Ely cathedral lantern.

Figures 141 a and b: The tendons of eight angelic hands hold the lantern of Ely cathedral.

An example (out of many) pictorial art works of hands in the gesture of holding a crown above Mary's head. Detail of Sandro Boticelli's *Madonna of the Magnificat*.

The hands that hold a crown gesture in religious and mythological art

Top: Detail of The Coronation of the Virgin by Raphael

Bottom: Detail of ceiling showing the Coronation of Venus at Charlottenburg Palace, Berlin.

Figures 142 and 143: The arch in iconography as a compartment for the body.

Left: Madonna, Reims 1211-1212. Romanesque trefoil frame for the head and shoulders, evokes the nave of a church.

Right: Madonna in the Reliquary of St. Ursula, occupies a Gothic elevation.

In this motif, an elevation similar to that of an arched nave is a frame for the sculpture of a Madonna. Its arch is Romanesque, and it displays a structural motif conceived as a frame around the Madonna that is formed by a taller semicircular arch and two flanking half arches. The carving of the arches above the Madonna's shoulders are slanted rather than straight, which makes them look organic. It could be seen as an aesthetic shell for the outline of the body. In the Christian religious texts the church is Christ's 'bride' and the Madonna is 'the Church'. Architecture has sometimes a close cognitive association with a shrine where The Mother resides.

Mosaic depicting an allegory of architecture in the nineteenth century Albert Memorial, London.

Figure 144: A much later secular mosaic also depicts human figures in the size of a building's central space.

Figure 145: Saint Nikolas, Luneburg, Germany, 1440. Flower-like star vaults.

Figure 146: Canterbury Cathedral Lantern. Fan Vaults.

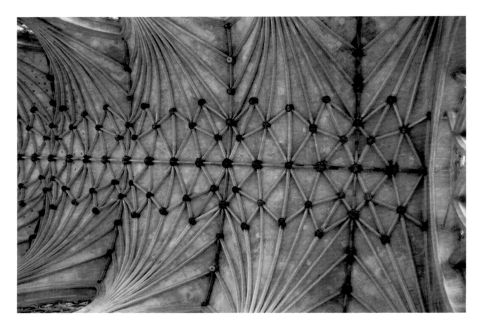

Figure 147: Ely Cathedral choir vault.

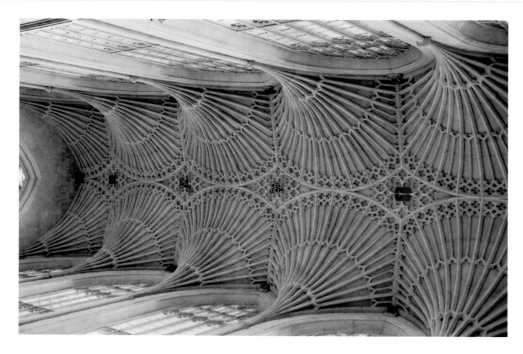

Figure 148: Bath Abbey Church fan vaults. The present church was founded after the Norman cathedral was in ruins in 1499. The abbey underwent major restorations in the 1860s.

Figure 149: Vaulting of Bishop West's chantry. Ely Cathedral.

Chapter 10

Vaults and Trefoils

Early vaulted underground spaces have been said to have provided the pre-literate mind with a suitable analogy of maternal shelter. There are pre-Roman vaulted cave chambers in Sardinia, Cyprus and Sicily, of early Paleolithic origins[136] that were shrines dedicated to a maternal goddess and have been said to symbolize a womb[137]. While the evocation of a womb in natural caves or built forms, may be a subconscious memory that all mankind share (a natural Jungian archetype), in the precarious lifestyle of pre-literate cultures, and under an early mode of communication, the analogies about the gestation of life were very likely consciously made. The idea that a cave type dwelling represents a womb converges with Buddhist thought. In Buddhism, the stone vaulted shrines that house a *stupa* or a Buddha, have also been identified as representations of a womb[138]. This association of the womb and underground or vaulted spaces is very likely part of an elemental sense of refuge. The association of a cave to the womb is manifested today in the metaphor that we sometimes read, that in death we return to the 'earth's womb', in which case the subconscious archetype is the personified concept *Mother* earth.

To a scholarly quest for the origins of forms, the fact that these vaulted chambers and other domical structures[139], consisting of underground tombs and holy baths, are earlier than structural vaults, is seen as providing ideas for building vaulted roofs[140]. This could be a reference to tunneled earth spaces (caves) lined with stones or bricks, which later might be

---

[136] The Mediterranean Paleolithic Age extends from 15,000 to 10,000 BC.

[137] Kaschnitz von Weinberg.

[138] Reproductive organs represented the power to generate life to early societies. For scholarly conjectures about design forms that follow reproductive organs see Kaschnitz von Weinberg.

[139] Small domical structures occur in many world regions and can be built in different materials, such as branches, stones, ice or mud. The Native American prototype based on branches may point to migration of this particular type. The fact that these examples are built for ritual reasons has helped to freeze its construction in time.

[140] Kaschnitz, 1944.

built in some arcuated method such as in corbels. It is no surprise to this study that a pre-existing form is important in some architecture, such as the case of the human head in the dome of the Roman Pantheon, and many religious buildings of several faiths, for example, that are designed after prescribed religious features. Also, to some extent, a shape could be chosen as a model-idea to be tested as a structure as in the inspiration of a niche in a shell. However, an architectural-scaled version of any test form, chosen as a model for a structure, still needs to offer constructional sense. With today's technology, an architect sculptor has freedom to achieve fanciful forms, but this wasn't possible in ancient times other than in a solid sculpture, as that of the Giza Sphinx. Thus, these processes are not as simple as their semantic expression is, and semantics can easily lead to categorize the physical reality of materials and construction and lead to imprecise understandings. Also, static processes generate shapes; structural forms; and it is easily observed that the psyche searches to find its symbols in the forms that have been made available in construction, just as it finds images in a sunset's clouds. Through embellishment and adaptations that don't affect a building's support, the symbolic recollection may become enhanced, as the Pantheon's symbolism was enhanced by coffering.

Among the structures that have become motifs that convey meaning, the vaults offer the highest clarity in their sheltering expression. The vault became more eloquent as it advanced toward the Middle Ages and did not become obscured by its transformation in time. The same applies to the vaults in miniature scale, or vaulted motifs in sculpture. This eloquence advanced in the minor crafts to a virtual obsession for vaulted motifs in the Middle Ages, that loudly celebrated architectural arches in carved stone. These miniatures celebrate the aesthetic achievement of the age's large functional vault, and the ability to create comes along with some conscious and subconscious suggestions, as both a divine canopy and a human-shaped refuge are represented here. The achievement reached in structural vaults was uplifting and breathtaking to builders and the medieval people. These miniatures can closely contour a human figure expressing silently the role of an architectural cover. They convey a sheltering notion that is expressed as a structure of complementary contour of the head and upper body (see *St Genis des Fontaines* frieze in Fig. 129 and also see in Fig. 1, the ivory figures underneath the dome's canopy, which are in individual niche-like spaces.) The Gothic vault was represented in a multitude of devotional sculpture in the church. In this miniature version the often-proposed natural association of a vaulted hollow space with a womb seems far in the distant past. Instead we see a representation of a built shelter as a personal scaled architectural cover for a standing adult human figure. The expression of this significant architectural meaning as a refuge is possibly a cognitive and sensory concept of a cover that is in the mind, as that of the Pantheon too; an outline of the brain's covering structures (or the upper body) as observed in some full-scale Gothic vaults. One of the most important roles of architectural embellishment seems fulfilled in this iconography.

The miniaturized depictions of buildings that were repeatedly placed above the figure of a person, indicated, on the one hand, the divine quality perceived in the structural full-scale vaults. The idea of divine refuge in cathedral art was comforting to an age that was concerned with hopes for salvation. The crafted small vaults and saints, evoke the sheltering role of architecture in a ritualized depiction above people's heads[141]. The saints under these aediculae express

---

[141] Lewis Kausel, 1982.

religious serenity or bliss. The motif of aedicule –shaped as architecture, i.e., roofing, arcades, monasteries, turreted vaulting– and a human figure is related to the protective niche. When these motifs contour the head and the shoulders, as in Gothic three-foliated windows it evokes a concept of a mantle or a hood. This is the trefoil arch[142]. The three-foliated arch is reproduced in windows, doorways, interior arcades, pulpits, the *Prie Dieu*, etc. in miniaturized and simplified form, in both stone and wood crafts, that include screens and pews (see Figs. 10, 142, 153 to 156). As conveyed visually in the Gothic sculptures, this trefoil is a fitting protection[143].

The elevation of the nave of churches (rather than the floor plan) causes a significant visual impression in viewers. These soaring religious spaces developed intense meaning and moved artisans to crystallize their arched configuration in sculpture. In these painted and sculpted representations, a church's side aisles can be portrayed with rows of a quarter rib (rather than semicircular ribs) at the aisles' ceilings. This probably seems logical, both as a frame surrounding a human figure, and in a structural scheme where a semi-circular arch needs lateral buttressing. Needless to say, when the nave's elevation is shown as a motif of a central arch with its side aisles, such as in altars and entrances, it is possible to shape the flanking smaller arches as only half arches so as to form a three-foliated arch, as it was done in the hammer-beam truss of fourteenth and fifteenth century England[144] (See Figs. 142 and 156). The parallel of nave's outline and upper body is probably arrived at as a fitting and harmonious frame or space, but not necessarily realized as a parallel of the body. The arch had approached a compelling form. This probably made the pre-Romanesque arched window, doorway and nave, fitting or appropriate to the mind, and this had a possibly subconscious force that gradually guided the nave to be transformed into the trefoil frame, that by joining the side aisles to the nave (as in Fig. 142, and in the structure of Fig. 156) it forms the outline of the human form. Sculptors multiplied the trefoil all over Europe and beyond, in miniatures and other objects. The cognitive evolution or progression of the trefoil, although unrealized, is about the relationship of architecture and the human form. It is an entirely cognitive and sensory expression. What it conveys silently is not distinct from the concept conveyed in several of the images shown in this study where architecture (or architecture-like head gear) surrounds the upper human form, except that the context, materials and the examples are different, but what architecture means or does to a human form remains the same. These observations may be further supported by other kinds of motifs as ancient as the years BC, or as far away as in India and South East Asia, in which the human figure fits within structures of small scale with trefoil outline (see Fig. 17).

As arches and vaults were the greatest achievements of their days, artists found them worthy for venerated sacred figures, and according to what the clergy wrote, these architectural achievements and their associated sculptures were thought to be the result of revelation. Again, this is important in the collective mind's ambit where these designs develop. Though not possible to ascertain, this conclusion from the part of medieval writers was probably derived

---

[142] A trefoil of three equal interlaced forms has had an old conscious or conventional symbolism as the concept of Trinity, which may have influenced the use and the spread of the architectural trefoil. The Trinity trefoil recurs in Christianity and other faiths.

[143] See Lewis Kausel, 2007.

[144] As it can be appreciated from Fig. 209, the interior elevation of a church, based on semi-circular nave and two flanking smaller arches, was already visualized in ancient Rome in the triumphal arch of Septimus Severus.

from interpretations of curious events and even dreams, hence, the medieval idea seems to agree with an involvement of intuition, in this perpetually-multiplied design of history.

As for the religious value of ritual imagery, it was created for belief-related objectives; for example, to feature portraits of saints, as a resource to those who need to pray for healing and favors, and also, for religious education, by identifying them, allegorically, through the themes they were known for. The constructed medieval church was a setting where all the persons portrayed could be encountered in spirit. It is well known that a religious building is a place to get in touch with the divine, and this is the notion that the rites emphasize still today.

CECILIA LEWIS KAUSEL

Figure 150: King David by Benedetto Antelami, ca. 1180–1190. West façade of the Fidenza Cathedral. Illustration by author.

Sculpture of King David found in the west façade of the *Fidenza* cathedral. The sculpture is in a niche and it has a miniaturized structural motif above King David's head. It is a case where an architectural configuration exists above human heads at two levels, and it informs three levels of architecture in miniature in one sculptural composition. The largest one is the enveloping niche (a domed architectural space in the manner of an exedra or booth) that shelters the figure of King David. The next in scale is the small configuration above King David's head. It is arched ritualized little architecture. The third level is the transformation of this arcade held by two columns, into three minor arches or three-foliated system that have been distorted to follow the contour of the heads in profile, of Joseph and Mary. The Child's arch is the one that reaches higher. It is in this motif where the distortion of architecture and its transformation into something that looks like a form of nature for the human outline is revealed most clearly.

The human figures show two royal persons, one of which may be a bishop. They are depicted in bending or sitting attitudes under a vaulted architectural tops, perhaps reflecting the shelter of the church as institution. *Illustration by author.*

Figure 151: Capital in the convent of Nazareth 12th C.

Figure 152: Jamb statues at Chartres Cathedral. Columnar biblical persons, Royal Portal. XII century. Illustration by author.

In the theme made of the two parts, the vault is small so as to be depicted in the scale of the head and conversely, the human figure is too large for a vault. But perhaps more important than the medieval language of the hierarchical dimensions –where clearly the saint is the important part of the motif– is the symbolic mutuality communicated visually by the two parts, by being shown as complementary forms of each other. The development of the Gothic architectural style took place in the twelfth century. As mentioned earlier in this study, this construction phase produced the acme of clarity of the vaulted sheltering motif in sculptural form. In its representation, the statue of a human figure is crowned with ease by arched complexes above their heads. The importance given to the treatment of the two parts in this imagery indicates a reciprocal relationship between mankind and building, and, between church (as God's protection) and the saint represented. There are examples showing an aedicule very similar to a crown. The symbolically-protective arches open in two planes. Some of the Chartres statues have been placed under these arcuated forms oriented in a way such, that a tip of a central articulating corner is right above the saint's nose. In these cases this corner is sculpted with a shorter tip. The morphological evocation seems to be one of an ancient helmet with a piece covering the nose between arched openings for the eyes contouring the eyebrows. This is a highly physiognomic evocation for this aedicule.

Figure 153: Timeless church detailing that shows the trefoil motif
that evokes the human figure.

Figure 154: *La Communion du Chevalier*, Reims Cathedral,
13th Century. *Illustration by author.*

Figure 155: Annunciation. *Annunziazione* Basilica. End of the thirteenth century. *Illustration by author.*

In the thirteenth century there are comparable motifs of arches that contour a human figure eloquently. Trefoil arches in niches, also known as Gothic cusped arches are used in a way that follows the head and the shoulders of the human figure. A match between the part sheltered (human figure) and the sheltering part (architecture) is present in this sculpture. The halos of both angel and Mary are of the same curvature as the framing arch. The two parts are complementary forms; a correspondence that involves the framing of the figure by the architectural interior, as a mold of the human outline.

Figure 156: *Misterium* by author shows the correlation of the trefoil arch and the human figure.

The Self (the subconscious idea of god in several faiths) may have been represented in the human outline of these trefoil structural arches (which are at the level of the ceiling. i.e., the mind) and in the central trefoil at the heart of the depiction. See Lewis Kausel, 2007 for more on this vision. *Original Artwork by Cecilia Lewis Kausel 1988.*

Part IV:
REPRESENTATION

Foreword to Part IV: Function and Image

In the view of this study, a vaulted design such as a cathedral's vault, may make progress in a sculptural direction in smaller versions of the vault and the form leaves its structural context. When we look at some examples, as the Gothic aedicule, we see that it displays interior and exterior elements from the full-scale architecture of the church (see Fig. 157). A structure like a Gothic arch and vault migrates from a cathedral structure to many other entities (aediculae, choir stalls, niches, furniture and head attire) and it hence transcends also various objects. We have mentioned too that the image of a medieval arch ends up as tracery, of smaller pointed, circular or horseshoe arches of varied sizes. These smaller structures are utilized as independent design patterns.

In other studies of the aedicula, such as the well-kown 'Heavenly Mansions' by Summerson, the focus is on the canopies, past and present, which is a category of artifacts based on a light type of cover with ceremonial function[145]. In religious iconography, an aedicula takes its configuration from various different structures, among which, common forms are a temple pediment, a church or monastery portico, a nave vault, an apse, roofing and covered palanquins. These are indeed very different designs. Instead, in this work, the focus is the Gothic aedicula as a miniature vault, or a vehicle that reflects the image of the architectural vault. The Gothic stalls, chairs, and other artifacts are also vehicles that display the configuration of arches and vaults. A vault is a visual image with an aesthetic power hence, it is useful to study architectural forms which have influenced the crafts, as resulting designs that have the capacity to become independent from functional architecture[146]. Our study has attempted to understand this visual phenomenon, whereby a static form behaves as a cultural concept, and by this we mean a design that comes to be identified by almost everybody. This visual phenomenon becomes firmly implanted in the memory of people.

We could look at a different example instead, such as the Thai temple and its associated images in Figs. 18a to 18d, and observe that the spire-like forms of the temple roof are found in the head attire of the temple's artistic human figures that are sculpted in several roles, including mythical animals, and they seem to have the objective of 'keepers' and 'guards' that dwell there, as well as brackets that support a cornice. In this exercise, we observe the utilization of design in sculptures, or better still, the meaning that architectural forms have acquired and how they are applied to the appearance of associated human figures. This cultural 'vision' of roofs is often important in the identification of design by culture and hence, for our exploration of the meaning of some designs, and therefore, their visual relationships. In the case of Western medieval architecture the migration of arches and vaults to the crafts took place in the guise of religious iconography.

---

[145] Summerson believed that the aedicule could have given rise to architecture. His own words on this matter are: "….the aedicule has been enlarged to human scale and then beyond" (Heavenly Mansions, 1963 page 4), and he refers to Gothic architecture as 'aedicular architecture of the grand order'. The aedicules, in general, can be observed to carry many varieties of designs, from a flat top, to a pedimental one, or from tent-like forms to umbrella and vaulted forms. The idea of aedicules as a category of objects is not tied to a specific structural shape or form; but it is basically any lightweight canopy. The artifactual features of an aedicule are ceremonial function, open typology and small scale. All such features define attributes that exist semi-independently from the vaulted image that the Gothic age used for the aedicule. In this current study, we look at the connection of the image of the Gothic aedicule with the vault.

[146] Lewis Kausel 1982.

A design concept that is abundantly reproduced is often shaped further and may become blended with some forms of imagination, showing transformation and sometimes distortions. In fact, a given old configuration that is visually well-known, may also be selected (sometimes in an unintended way) as a new meaningful motif. It may be streamlined or reshaped successively, and it may begin to sometimes reflect an 'animation' (possibly a tendency to a humanization of design, or as we have seen in the case of supports, a representation of a metamorphosis between animals or semi-human figures and supports, and the human outline of the vaults.) Vaults and domes show instances of transformation into bishop miters and tiaras in objects; then pillars can show a transmutation into goddesses, and herms, or Thai temples into sentinels.

Figure 157: The aedicule as a decorative miniature vaulted space with ribs, gothic, lobed and trefoil arches.

The medieval carver seemingly delighted in reproducing arches and ribs of varied forms.

Above: This is one of the many Gothic aediculae with a vault's ribs and rosettes at Ely cathedral, England, at Bishop West's chantry; began in 1520. A chantry is a chapel within a church, endowed for the signing of masses after a founder's death. Almost invariably these chapels have a screen with many carvings and sometimes the grave of the bishop responsible for its construction. This is not the only chantry of Ely cathedral. Opposite to this chapel there is Bishop Alcock's chantry built in the 1480s.

Chapter 11

The Import of Representation

A creative exercise that may elicit a particular emotional response is the depiction of specialized objects in a miniaturized scale. This is done for aesthetic and other visual ends. The miniaturization of a specialized form is the most common example of scale change, but there are also occasional instances of small items that are shown in enlarged scales. Miniaturization depicts an originally large-scale functional structure in a small dimension, and it can be complex as if in full detail allowing viewers to hold, for example, in their hands, a tiny building, which is essentially a reproduction of a large monument. The form of a miniature is most of the times a case of representation. Through miniaturization a stone structure, such as a temple or church, becomes an iconic gem and a sculptural depiction of an intellectual achievement. The idea of executing an attractive building, or a ship or other technical form, as a scaled tiny object reveals to this study that the forms depicted in miniatures have captivated the intellect. The miniaturization exercise involves sometimes a desire to study a structure by replicating it. It also may involve the mind's vision and wish to execute a structure in more precious materials, such as silver, enameled metal, ivory and precious stones as seen in history's reliquaries and silver ships. From simple observation of represented structures, the psyche's inner vision of architecture seems more precious than that of a prototype building; hence the crafts interpret a prototype edifice in special materials. The miniature elicits wonder in viewers, and admiration for the meticulous mind that took the time and effort to come up with the conception (see for example, Figs. 1 and 158).

As suggested above, a change of scale can also be implemented after a wish to enlarge a normally small object, such as a key to a treasure or city, or to other important door, in a much larger scale and finer material (see Fig. 159). A key is a functional artifact that is highly symbolic in culture[147], and perhaps for this reason some individuals would enjoy ordering a very large key to be displayed at a reception room in some institution. A key sculpture is also a fine gift for someone with a guardian role. A finer material than the commonly used in a key, especially a precious one like porcelain, enhances the aesthetics of a common key, especially that of the configuration

---

[147] Sufficient should be to notice that the apostle Peter is identified by holding a key to Heaven.

that opens a door which can be designed with geometries that attract the mind. Though the change of scale from small to large is uncommon in design, it too shows us imagination at work, just as the reliquary does, in the idea that a key can be an aesthetic form. The key certainly enters a different domain by this kind or re-interpretation. It becomes aesthetic and interesting, not to mention that it becomes a rare item and a curiosity, worthy of being displayed.

Focusing now on miniatures, the reliquary of Fig. 1, shaped as a preciously enameled church, may be first perceived as a container for a church's treasures, but learning about its background reveals that its treasure is not made up of jewels exactly. It is actually a funerary urn created to house the relics of a saint. Discovering this function changes our first impression of it as a jewelry case to its somber reality as a cinerarium. We may empathize with the concern of the medieval period for miraculous cures and solution to problems, and at the same time understand that its treasure is substantially more valuable than diamonds and gold. The reliquary is indeed an emotive container that allowed the church to present these relics to people who needed divine intervention. The reliquary is one more design that reflects the pathos of the human condition that can affect people sometimes. Indeed before the modern age's medicine and care for citizens, this pathos must have been common. The respect of the faithful for the remains of a saint required a precious container made with great care for its presentation. Comparable motivations surely existed behind the crafted cinerary urns of Rome, shaped as vases, sculpted sarcophagi, small domes and portraits of deceased relatives. In the faith of a believer a dead person can somehow still be present in his/her remains, which in the case of a saint, are sacrosanct. Hence, the ciborium is a reverent little place of abode -sometimes with a somewhat perceived parallel function to that of real architecture. Its semblance needed to be dignified by the sacred building it displayed and hence, it could be honorably carried in processions[148] making this disturbance of a dead saint, socially acceptable. This much and maybe more, is expressed in a 'silent' or visual way by the associative connotations of the image utilized in this container.

The miniaturization of a building is an ancient craft, and it is not always a scaled model. Miniaturization is indeed an idea that arises in the creative activities of artisans, even though from there it can advance to a fine craft and eventually end up in a museum collection's treasures. Inasmuch as the activity of reproducing architecture in a craft reflects that the building means something, it is significant to our knowledge as to why do we shape architecture and enhance buildings and artifacts. The terra-cotta miniature temple shown in Fig. 158 is a fitting ancient Roman example of a meaningful form, a temple. This miniature allowed people to possess, carry and display in an interior space, a crafted building in portable dimensions. Today, this ancient idea may still be present in a neglected craft, the souvenir, which is often a crudely-crafted miniature that is mass-produced. The souvenir is a type of figure that does what a post card does, except that it is a three-dimensional form. The same interest that makes us photograph a building can also make us acquire one of these souvenirs. Miniature replicas are molded in economic resin today and are manufactured in varying sizes. There is no craftsmanship in such economical objects. A common example is the tower of Pisa souvenir. What is the objective of an economic likeness of the tower of Pisa? It allows all visitors to take along something from the site that can be purchased for little money. It recalls Italy through the white tower whose precarious inclination moves the public interest, while it holds to the ground without collapsing.

---

[148] Miniaturization responds to more than these consecrated functions. It also reflects the pleasure of possessing scaled aesthetic technological structures and other rare artifacts that have been executed in exquisite materials, as for example, in silver or brass or in real materials.

classicism that became common in neoclassical design. Perhaps some of the various and very ancient circular temples of Greece, may be the model behind the Lysicrates cylinder monument. These and other possible extant forms of the most ancient world regions indicate that the representation of structures seems to have accompanied civilized mankind all along. Then, there are other types of representations that picture structures in miniature, such as the end stones of the aqueduct of Segovia that depict this public work. This portrayal was executed at a point in time when the structure was clearly esteemed, not only for its service, but for its looks in the landscape. These representations of the aqueduct were added in the XVII century and reflect a dedicatory intent, or the city's recognition of the structure.

In the Pompeian altar shown in Fig. 168 a gabled temple portico is miniaturized and has become a frame for the scene depicted that shows *Lares* and the *genius,* the protectors of a household. What these portrayals of a temple facade convey must be a temple's identity, as it relates to the concept of sanctuary and the dwelling of gods. However, once this façade is represented in a furniture piece or an interior (as in Fig. 169) it is no longer a religious façade, not since other religions replaced ancient mythology, but it has retained its distinction as it is passed to the minds of later users and other cultures. It is a prototype image that is usually perceived as a distinctive entrance or simply a form that gives classical style to a wall area and emphasizes its elegance, as a chimney mantel, a presentation for an important entrance to a room or area on a wall or fireplace.

History's builders often created a small model that portrayed a proposed building project to a patron (see Figs. 170 to 172). These small models have been interpreted as gifts in the case of the ancient Roman temple of Alatri at *Satricum* (Fig. 158). In medieval to Renaissance examples building models were executed for church projects to be built. Probably they were made to show how a structure is expected to look like when it is built. The models depicted in these figures are evidently not test structures but most are executed from structures that are already possible architecturally at the time[156]. We see these models abundantly represented in medieval art where a patron, king or bishop, offers a temple to a holy person. They commemorate the donation or the dedication of a particular church or cathedral to Mary or to a city, and may include the donors who offered it, or a bishop who ordered it. These depictions of miniaturized buildings are highly repetitive, thus formalizing the idea of a miniature building. This may take the curious form of Fig. 171 where a bishop seems to display five churches associated to him (for which he may have been responsible). The model tied to his head by his miter ribbons, could be the capital church in the group, from its position in the representation, or one that has the highest or original significance.

There are murals in Pompeii that represent outdoor scenes of buildings and streets painted on a closed wall. The lack of windows in these interiors probably compelled a representation of an open urban scene, as an illusion of a desirable condition. Comparable solutions may be the balconies[157]

[156] We must clarify that this work in no part or text has implied that a symbolic image, nor even a semblance or concept can lead to the solving of a static system. This would misunderstand the reality behind figuring out structures. We hope that this work can provide good enough examples of the domain of inspiration and its restricted niche of influence over structural design, so as to make this point as clear as possible.

[157] Intuitively and functionally, a balcony is a space higher than floor level, to look out to passer bys, or down to a city, to a crowd gathered below, or to the distance. It is used as a place where a crowd of people can see a dignitary, such as in a ritual papal blessing. A balcony responds to the wish for fresh air and to social functions. If a balcony doesn't function it may still enhance the reputation of a limited or modest place. The nonfunctional roof railing or widow's walk commemorates a regional US custom of colonial women looking into the ocean waiting for their sailing husbands.

that don't function, roof railings that are not part of a terrace, doors or windows that are not meant to open, and hundreds of other examples, just as the *Mastaba* tomb door that is only an indication rather than an opening. Representation thus, helps the psyche obtain a more complete image of what it would prefer to have. Whereas some of these semblances relate to a desire to compensate some absent feature that the psyche needs, other times a nonfunctional form stands for something previously functional and a concept ghost, which in the perception of a designer, must be present because it adds a wanted character or the necessary vitality that the object type displayed before[158].

One of the mind's most critical specializations is the ability to reach other minds. This leads us to visual communication. The semblance of structural support in artifacts, for example, seems to be an aesthetic or symbolic representation that fulfills what 'should be fitting' in it. That fitness is one of those aspects of design about which there is wide agreement. The example of a plain furniture leg, that 'begs' for some fluting in order to look 'right' is a treatment evoking a semblance that seems 'fitting' to the furniture part[159]. From the examples studied, it is important to conclude that designing for consumers involves giving a semblance to a projected entity (and many times even a wanted outward show as the connotations of this phrase is currently understood in the US). The semblance of design has up to our days come from methods, imagined sources, or has been inspired by existing forms[160]. Semblances involve mimesis and ingenuity. Distinction, elegance, reputation, status or other necessary characters can be informed by the semblance created or selected. Semblances have been achieved in design either without masking the framework or constitution of an artifact, or, masking it partially or entirely, but masterfully, as in the exceptionally good ornate examples presented herein. The most vital quality of the outcomes of design are not so much how much, or how little, treatment masks a framework, but whether the semblances created are fitting to the searches and sensitivities of the minds that perceive them and enjoy to experience them. These sensitivities take care of arriving at quality and fit in all design.

Representation certainly is not absent in most modern design, even if the early understanding of modernism came with a resolution to advocate abstract forms, and abstract examples were indeed designed. The reality of many works is however different in some ways from what it has been advocated, which happens when a style of design is applied by thousands of designers, each one with an independent and creative mind. Sufficient is to cite the evocation of buildings and products of industrial nature in modernist design (as for instance, the sometimes called "streamlined *moderne*" that was sometimes shaped to evoke subtly a steam boat's windows and chimney; or a jukebox or nickelodeon; or the late twentieth century cylindrical borer or grain elevators. These are all inspired forms, even though many were designed to represent the abstract aesthetics of the twentieth century, and these cases must also be acknowledged as design with

---

[158] The case of the Roman monumental arch 'needing' the post and lintel Greek façade.

[159] Alternatively a furniture leg may bring to mind the idea that it 'begs' for an animal leg.

[160] In the Carthusian Baroque design of Spain, interiors fulfill imaginary circumstances that dematerialize the architectural surfaces into other impressions. In La Cartuja of Granada, design is masterful in the representation of an imagined white textured dimension, of perhaps clouds (shown in white plaster). This immaterial suggestion in plaster is observed above the wainscots disposed as pedestals under these 'cloud walls". It is a composition where white ethereal surfaces are the focal point. Above these clouds is a representation of a vaulted and another different dimension: Heaven. The profuse plaster decoration dematerializes the middle walls into ethereal elements: air and clouds, which were the space that leads to heaven to the historical mind. Among the most impacting designs of history, space was animated in these ways to impart suggestive power to static objects. This is what decorative design tried to attain.

a tendency toward the figurative if we want to be accurate. In the end these formulations are a matter of interpretation; this is true; however inspiration is part of a recognizable dynamics of emulation in the arts and hence, of design perpetuation[161]. Inspiration has also focused on factories and imagined future societies of industrial workers.

As to the connotations that some images can suggest, that are used in representation, it is important to understand that in culture a single form may communicate dual or even multiple meanings. In fact, one and the same form can carry not only more than one understanding by viewers, but both realized and unrealized –or subliminal– evocations. The concerns of society are frequently reflected in the production of a time. It is possible that some viewers one day can come to 'see' clearly the connotations of a form, even if they didn't notice it before that momentous day. A subject matter or a design that carries a meaningful concept indicates something about its suggestion to many minds, along with its character (that can even be pathos), in an unspoken mode. As long as the form in question exists, its visual suggestion can reach receptive viewers despite changes in aesthetics in the arts and design; thus, people of recent times still cherish such meaningful forms of the distant past[162]. To assist the idea that a design may affect our perception at more than one level, we can cite the meaningful mythical images that actually function at subliminal levels[163]. Now, what is the import of representation in culture? It is certainly the vast world of suggestion in the production of material culture.

---

[161] The emulation of modernist design focused both on designs by the modernist masters and the products of technology associated to the new pleasures of the age, such the aerodynamic forms of airplanes and automobiles.

[162] Lewis Kausel, 1982.

[163] A parenthesis illustrating a concept with many symbols may clarify how multiple meanings are possible. There is an ancient product of imagination that takes the form of a large reptilian creature, sometimes as a snake and sometimes a quadruped and winged dragon. This reptilian form surfaced out in legend and religion where it blended with allegories. This form is said to be an archetype even though it no longer seems to affect contemporary people. Perhaps the frequent depiction of dinosaurs in film and video games reptilean creatures in our times provides a 'benign' outlet for the reptile archetype making it extinct and unreal. But a glance at old art and coats of arms corroborates that dragons and snakes have inspired respect, if not fear. In some other usages they represent wisdom and luck, and in ancient Greece and Rome, a snake represented the *genius* and accompanied portrayals of Hygieia, goddess of health. If we come across a snake unexpectedly, it can make us shudder and put us in a warning state. This is a spontaneous reaction that is likely hard-wired and it is not allegorical. Perhaps from this reaction everything else around the connotations of this creature derived. In the Scriptures a snake symbolizes a boding evil that sometimes affects destiny. The reptile was perceived as lying in waiting for an opportunity to influence people to sin. It was subjugated by grace, thus it is depicted under the feet of the Virgin (the protecting Mother), or surrounding Saint Ursula without harming her. To C. G. Jung the biblical snake and other associated reptiles and aquatic forms represent our own unconscious. This interpretation also occurs in ancient India. The snake also carries in a cryptic way the connotation of knowledge (e.g., the snake of the Tree of Knowledge), which adds meaning to the same archetype. In epic legend dragons were dangerous angry creatures or a wizard's obstacle in a journey. Saints and heroes conquered evil as in Saint George killing the dragon. However, in the Far East a dragon is a representation of the king, and it provides protection and good luck. Clearly then, reptilian imagery has quite a number of symbolic readings and we have discerned only some of them. The possibility of multiple readings should be expected too in other imagery that becomes meaningful.

Representation

Figure 158: Miniature Temple of Ancient Roman Times.

Small temples are thought to have been given as presents for consecration and inauguration rites. The material is terra-cotta. This particular example was found in the ruins of the *Satricum* Temple, south of Rome. It is believed to represent a temple built in 400 B.C. identified as *Alatri* by Kaschnitz.

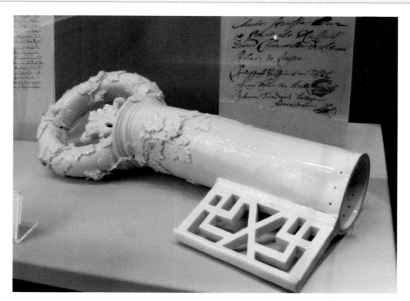

Ursula Zastrau

Figure 159: Oversized aesthetic porcelain key.

Meissen porcelain factory display.

(a)   (b)

Figures 160 a and b: The goddess Athena Parthenos, compared
to classical column of Paestum.

Athena Parthenos pensive, looks at a stone slab.

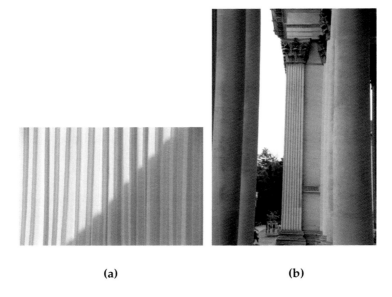

(a)   (b)

Figures 161 a and b: The Fluted pilaster of a classical building and drapes.

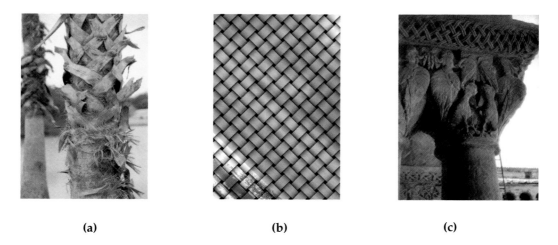

(a)   (b)   (c)

Figures 162 a , b. and c. Overlapped palm bark is similarly entwined in nature to the plain weave knit.

Right: Romanesque capitals of the Monastery of Silos in Burgos, Spain, show a basket weave motif.

Semper observed in the 19th century that several ancient ornaments arose in the technical crafts, and became important to a nation. Hence, when the same society developed architecture, it represented its ornaments in buildings.

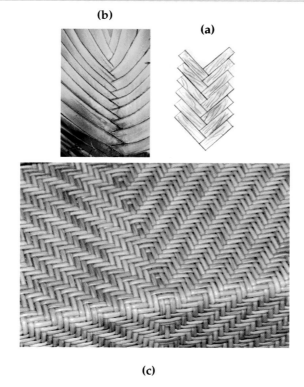

(b)

(a)

(c)

Figures 163 a, b and c: Pattern of palm leaves is somewhat similar to the herringbone flooring patterns and to a weave used in wicker.

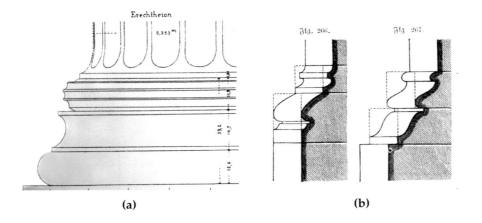

(a)

(b)

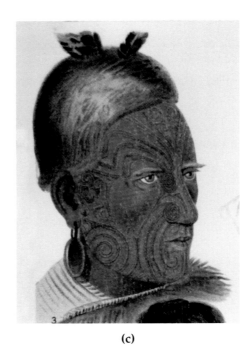

(c)

Figures 164 a, b and c: Classical molding at the base of a column shows adaptation to weight. Column base from the Erechteon. *Source: (a) Left: von Egle; shaded by author. (b) Right: Warth. (c) Illustration from Meyer's Lexikon.*

Figure: Some linear decorative patterns around building's surface openings may be assocaiated to a visual logic related to physiognomy, as in facial design that enhanced expression.

Left: *Germano-Romisches Museum.* Cologne. Author's sketch.

Right: Detail of frieze shows wicker chair. Roman tomb. *Landesmuseum,* Trier.

(a)   (b)

Figures 165 a and b: Limestone sculpture depicts a wicker chair that evokes a user.

Ornamental classical columns. Engaged columns evoke architecture in a stone cylinder.

Wedgewood vase 1780–1800 evokes the Lysicrates monument to convey Neoclassicism in the vase. Also shows female figure in pensive or funerary attitude.

(a)   (b)

Figures 166 a, b, and c Overleaf: Choragic Monument to Lysicrates, Athens, c. 335 BC. Urban sculpture. It incorporates the images of columns and entablature. Photo released by author Greenshed into Public Domain. Wikipedia.

Schinkel's analogy to Lysicrates' monument

(c)

The top part above the roof of the garden structure *Grosses Neugirde* (Great Curiosity) that Schinkel designed at the Glienicke Palace, Potsdam 1935–37.

Figure 167: Marble frieze slab displaying a circular temple of ancient Rome, perhaps a temple of Vesta.

It possibly depicts Augustus' Temple on the Palatine Hill. The relief shows a temple that had windows screened by stone tracery. This tracery was probably attached to stone posts behind the columns as it would have been more difficult to attach these screens to a column shaft. If such was the case, the columns we see were a surface image, or engaged columns. The bas relief shows incorrect perspective as it is visible at the stairs, doors and the temple's body. The rough sketch follows the original.

Vesta was the Roman goddess of the hearth and was associated to a permanently burning hearth flame that was tended publicly by vestal virgins. The temples of Vesta were circular in plan, in honor of the ancient Roman round hut, which had this disposition because of being arranged around the hearth. The shrine of Vesta was located in the Roman Forum and it was of great antiquity.

Figure 168: Altar in Pompei with pediment and fluted columns.

Altar of the *Lares* depicting two *Lares* on either side of the *Genius*. AD 69–72. House of the Vettii, Pompeii.

In Roman religion the *Genius* is "the begetter" the procreative power that enables a family to be carried on. In its earliest cult it could be the genius of the Roman housefather or the *juno* of the housemother. The evidence suggests that they were male and female forms of the family's (or clan's) power of continuing itself by reproduction. The genius passed at the death of the head of the household to the successor. It was often represented as a snake, although it was also represented as a young male engaged in sacrificing. The *Lares* were originally regarded as divine ancestors. The *Lares* were worshipped wherever properties adjoined and inside homes their statuettes were placed in a *lararium*, a domestic shrine. The genius came to be thought of as a guardian angel that each individual worshipped on his or her birthday.

(a)

(b)

Figures 169 a and b: Classical interiors commonly apply temple
facade imagery to frame doorways and wall areas.

Top: Pediments and columns in wood. Schwerin Palace, Germany.
Bottom: Ionic pilasters and cornice at a Tudor and Jacobean revival
mansion. Fireplace at Shaw Hall, Mount Ida college, Newton, MA.

Figure 170: Detail of Mosaic shows Justinian presenting a church
with an umbrella dome to Mary and child.

The iconography places the bishop in a central
position relative to these five churches and seems
to depict the Archbishop of Cologne. Several of
the domes seem to have the 'umbrella' typology
(see Fig. 1 for an example). The model tied to the
bishop's head by ribbons seems to depict three
onion domes (although this type is commonly
reported as having developed in Europe, in the
16th century.)

Figure 171: A bishop with a nimbus and five church models. 1056–75. *Illumination is at the
University Library of Darmstadt, Germany.*

Figure 172: Michelangelo Presents his Model for the Basilica of Saint Peter.

Chapter 12

Ornamental Architectural Images in Exterior and Interior Applications

This section discusses further some of the purposes of columns that don't support; doors that don't open; depictions of architectural columns and windows where there are none, and niches. There are cases of blocked arches and porticoes, windows and balconies that do not let light or air into buildings, nor do they allow people to see outside (or be seen from outside), but they serve the enigmatic purposes of pure representation.

Stylistic application is substantially different from a pictorial portrayal of a structure as in landscape painting, and this is an interesting aspect of the perception of that domain of design we identify as style. Painting has another type of intent and inspiration than a stylistic syntax which has very particular visual purposes. Stylistic application of visual motifs gives character to a new creation, along with the many other aspects of design migration mentioned so far. The logic of perception makes these two types of representations necessary and different from each other.

Of the early architectural parts to be applied without a practical sense, the representation of an entrance in a *Mastaba* tomb of ancient Egypt can be cited. This indication of a door concept by means of a recess was not meant to be opened. This idea is intended to create an impression of a door. A door that doesn't lead anywhere is designed to satisfy the expectations of someone who wishes to see at least a representation of an access in a tomb, as this is surely a more dignified design than one that ignores a soul's 'wish to move freely'. It has been said that this representation allows the soul or Ka to return to the body. The *Mastaba's* blocked entrance was in place sometime in 2700 to 3200 B.C., letting us understand that these impractical representations have been needed for long. Such designs are not meant to be deceptive, and this detail is important to our judgment of design. There are no 'good' or real design solutions for ritual needs other than satisfy them, even with an unnecessary or a non-functional design.

Concentrating next on particular structural elements, we observe reapplied images from Greek structures in extant sarcophagi and other smaller artifacts from Etruria and Rome. The Italian

Renaissance reused these forms as well. The niche, the dome, the pediment, the column with capital, the pitched roof, the Ionic volutes, the dentils, the 'egg and dart' motif and many others, were depicted in tombs and altars. This is done for the cognitive reasons[164] that fall within the larger understanding of our term 'representation' (see Figs. 173 and 174). A connection between architecture and the last abode appears expressed in different ways in design. The ritual nature of temple prototypes had something to do with it, and likewise the ancient customs of providing a house with menagerie for the dead; but architecture also has aesthetic recollections sometimes[165] whose impact is curiously of a funereal type of beauty. Aesthetics and the sadness of loss can blend together in some historical buildings, maybe as they can blend in some historical theater pieces and other art too.

There are in design, then, practical and non practical reapplications of columns as pilasters, for example, in interiors. Decorative elements are widely used to modulate interior elements and subdivide wall areas, and as mentioned in moldings and coves, to mask joints in wood and to regularize minor defects. The refinement of surfaces to produce unflawed results is no doubt an essential practical reason for embellishment, i.e., a representation of a motif that can hide a flaw. The ancient temple *metopes* with friezes were applied to cover the joint of two parts that are behind the frieze. Then again, a functional chair rail prevents damage from wall nicks from moving chairs against the wall, but this railing can be attractive and feature a nice molding. Then, interior wooden wainscots approximate a sitting height or a mantelpiece height, or even a paneling elevation. Wood paneling creates warmer interiors than those in bare stone finish. But the use of non-functional structural elements also responds to visual expectations, such as interior hierarchies, when column-delimited areas are built around a door, or to highlight art work, or define a division from a vestibule to a living room.

The custom of using every wall surface to represent squares and rectangles that are molded or indented has sometimes been thought to reflect a tendency to transform walls into a richly decorative outcome. The reference of Mario Praz *horror vacui,* has been mentioned in the design literature as a human tendency to fill plain space with decoration[166]. *Horror vacui* is a phobia of empty surfaces reported in unstable psychiatric patients; however, in the arts it is a normal habit to fill in space and it was widespread in the past, being more intensely practiced under certain periods that considered decoration highly aesthetic, such as the Islamic and the Baroque. Complex decoration that is aesthetically-done stimulates the mind. For example, it provides detailed gear to our drive to surround ourselves with things that look fantastic and luxurious. A highly complex design can also stimulate us to find in a configuration what we store in our mythological mind, or the 'forms in the clouds'. In order to achieve textural effects on large walls through minute decoration, craftsmen reproduced in a 'slavish' way the same tiny forms. This was meant to satisfy the desire for luxury of patrons.

---

[164] In fact in Etruria, the sarcophagus of *Cerveteri*, from 500 BC, displays Ionic scrolls above a vertical support (under the sculptured torsos of a recumbent couple.)

[165] See for example the death face with a terror expression surmounted by a shell in the Renaissance pedestal of Fig. 105, in between the two ram heads. It reminds of the ancient association of the protective shell with corpses. The shell is also represented in the niche of the same pedestal —which in this case is a commemoration of the ancient architectural niche. We have introduced the mausoleum-like quality of a low dome as in the Pantheon, the empty helmets and mourning sculpture of classicism.

[166] By Gombrich.

With regard to frames, the public relates with ease to a display of framed scenes, as artwork and photographs, showing people, places and time periods. In framed accessories the concept of doorway that we have mentioned, reflects transition from one delimited environment to another (as in one dimension of being to another.) The idea of a window, as indicated by framed photos and art, may create a recollection of a window to a different place and conditions other than that of the room's ambit. Additionally to this objective, the framing of interiors by pilasters and columns are many times related to the definition of smaller intimate spaces redefined within large rooms, examples of which are alcoves, niches, recesses and nooks that can be indicated visually by pilaster-framed subdivisions. There are marble slabs from Greece that show early applications of the architectural motif of column and pediment as a frame[167]. Rome adopted this form for domestic altars (aediculae), such as tabernacles and *sacraria* (see the altar of the *Lares* of Pompei in Fig. 168), and in canopies for urban statues. A Roman bas relief also shows the head of Protesilaus framed by two miniature pillars and a pediment. These depictions may have caused the post-and-lintel façade and pediment to be perceived within a physiognomic logic at some point, and from here it passed to the frontal area of helmets[168].

Interior millwork subdivisions with structural-looking frames and elements obviously convey classical style to a room. Similarly, wherever compartments, wide or narrow, are desired, the use of dado panels (whose appearance in very ancient temples indicates a likely derivation from column pedestals) also subdivides wall lengths and evokes classicism. For example, see the interior of the Pantheon under the coffering where there is a belt of window-like aediculae (added in the Renaissance) and the pedimented niches on the first floor that evoke classical temple porticoes. They are not real windows or porticoes however, and they are embellishments of wall areas. Also see the Pompeiian altar of Fig. 168 that similarly mimics a temple facade. They are cases of migration of temple facades as ornament. In terms of early ideas for the reapplication of temple facades in interiors, some representations of urban scenes in wall paintings in the walls of Pompeii may show a long-standing practice of portraying architectural murals. Some Pompeian murals were painted after theater stage designs. See also the Florentine Renaissance *secretaire*'s décor of Fig. 7 that looks like a multiplication of miniature stage backdrops.) The ancient theater led G. Semper in the nineteenth century to theorize that architecture ultimately developed from the theater[169]. In a not too different general idea, the theater of the Middle Ages continued to construct scenes bounded by architectural frames that were meant to represent built spaces[170]. They were termed mansions, a Latin word that Rome had used in a different way. Later, in the Renaissance, non-supportive interior pilasters and bases became fashionable and this custom lived on well past the Renaissance.

---

[167] A repast scene in A. Rich, 1893.

[168] See discussion about the exterior of the Roman Pantheon in Section 4.

[169] The theater relies on mimesis (representation) and historical theater is additionally thematic. If this derivation of architecture was correct, evidently the tendencies toward decorative mimesis in architecture and its thematic building prototypes would be an offshoot of the theater. Mimesis and themes are products of the collective mind, and it is here where we can find the similar tendencies of buildings and stages.

[170] The most interesting examples depicting architectural compartments are observed in religious iconography. These depictions of compartments try to indicate notions and connections across different eras, or the ideas of religious prophesy and precedents with their fulfillment in later ages. Iconography tries to convey these ideas in part visually and in part using an organizational logic. It is necessary to know the story to begin to understand the meaning of the themes depicted in these compartments.

In the case of a domical niche, a sculptor could become inspired by an apse and replicate the apse in a niche, thus passing the structural form to the niche. A niche would be a miniaturization of an apse. However, there are reasons to think that a niche portraying a scallop on top was created before the scalloped apse for the objective of producing a small domical model or test form, that in a way 'gave rise' to the apse. Though impossible to tell today, the widespread domical niche could be a long-standing emulation of a test structure: the ancient niche with a scallop shell. The Florentine pedestal that supports the 1554 statue of Perseus (see Fig. 105) indicates a belief that a shelled niche was built interiorly with concave roof tiles. If this was done in ancient times, the larger scalloped semi-domes of ancient Rome which are mentioned by Hersey, 1937[171] could be its offspring (see also the shell niches on Figs. 188 to 104 in Section 18.) The shell is believed to be behind the design of scalloped arches in Spain[172]. In such cases, the architectural structure of the niche was aided by a form of nature executed in masonry. An extant Baroque chair of Switzerland features an exquisitely carved shelled niche (see Fig. 8) which shows the continued repetition (or commemoration) of the ancient architectural niche, even in artifacts. There are also innumerable fountains and altars as well as chest-of-drawers of even later centuries, such as Queen Anne and English colonial examples in the US, that feature several miniature shelled niches on a high drawer. In American Victorian interiors there were panels above fireplaces with a finely carved shell pattern. At any rate, both a frame of pilasters –with or without pediment– and a niche establish a small space in a large room. Both frame and niche are associated to a more intimate space than that of a room; perhaps more fitting to the human form than the actual large practical spaces. Beyond possibly being a test structure, a niche in an interior is an architectural compartment built in the same scale of a statue, meant for its space, hence, a niche is an analogy of architecture in human scale[173]. Evidently a niche is functional[174], as it was used to shelter a speaker or to protect fine art, preventing damage from the traffic of people in public buildings. However, a niche also depicts divine refuge in a sanctuary for religious art, not to mention the representations of holy cities by the miniature forms above Romanesque niches (see Fig. 150). The refuge of a small architectural space is an interesting cognitive concept that was discussed in the trefoil motif of the arch in Part III. Sometimes a niche is left empty in architecture and it may be waiting to be occupied by a sculpture, or may speak of the commemoration of the niche itself. The shell niche has enjoyed favor as an architectural motif as observed in the migration of this form to artifacts (see Fig. 8) but a shell is also a natural form surrounded by symbolism worldwide, that shows to all peoples a functional protection, in particular of a pearl (a treasure)[175] in the oyster, hence it is one of nature's 'architecture'. The scallop shell has been depicted in some niches in ways that are also reminiscent of the brain and the skull[176]. The endless worldwide depiction of the scallop shell and the conch shell seem to say that they are archetypal forms. In an early application in Rome, we see it connected to the design of a mausoleum (see Fig. 188) possibly because a shell also symbolizes rebirth in most cultures.

---

[171] Including the skull, see Lewis Kausel 2007.
[172] Torres Balbas.
[173] Mostly so, though because in classicism everything has to be proportioned according to set rules, in a large space, such as the basilica of Saint Peter's in the Vatican, niches and statues are very large.
[174] The niche existed as a rectangular wall recess and was in place in ancient Greece, in the Acropolis in Athens. It served as a retreat for meditation or a place of discussion for poets and scholars.
[175] Eliade has lengthy accounts of beliefs associated to shells in different world regions.
[176] See Lewis Kausel 2007.

The ancient connection of architectural forms, sarcophagi and altars.

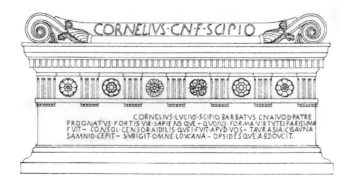

Figure 173: Sarcophagus with temple tryglyphs, guttae, dentils and scrolls. A wish for a dignified resting place is possibly what links architecture with funerary solemnity. *Image source: meyer's Lexicon.*

Tombs were pre-existing to the creation of many structures[177], however, the form of structures still migrated to tombs, when prototype buildings were in existence and became common images. The house and the grave offered analogies between them, and so did temple and mausoleum, both of which had religious conceptual foundations.

The arches tend to surround the human heads as it has been seen in other sarcophagi. The scale of the arches is too small for the scale of the persons portrayed. They fit tightly in the dimension of these arches, so they are ritualized as compartments or niches. They are an iconographical representation of the world of people and their buildings.

Figure 174: Christian Roman Sarcophagus with arcades, pediment and acroteria.

---

[177] Some basic architectural forms may have been first tried in tombs.

Chapter 13

Structural Ornament in Furniture and Other Crafts

The idea to portray building forms in furniture could have come up in the interior trades. Furniture is a functional type of artifact whose embellishment was important in history. Appearance is still important to the users of these serviceable items, in particular in pieces assigned for living and reception spaces, where furniture plays a role in determining the refinement level and distinction of a place. A thought that chairs are not really essential to mankind and that their role is primarily social, has been mentioned[178] based on the fact that some complex cultures didn't use chairs in historical times, and this fact seems to strengthen their representational role over their function, in the writings of some furniture historians, even though they are so essential today.

Concerning how the idea to utilize a structural form in furniture may have come to mind, the following hypothesis seems plausible. A cabinetmaker infuses ritualized building elements to a piece or object, so as to give a particular distinction to it, and so that it should harmonize with an interior[179] (see Figs 1, 3, 4, 6, 7 to 13). Some famous excellent architects and designers from the eighteenth century to today have required that furniture pieces be suitable for their architecture and they designed interiors themselves, such as Robert Adam, William Kent, Schinkel, Mackintosh, Van de Velde and the twentieth century chairs of famous modernist architects. In the case of the same mind behind the design of both building and its furniture, a migration of building character to the piece seems straightforward. However, the influence of architecture on furniture is ancient, as the influence of roof tops and even column capitals on head attire is too, occurring in classicism, and so is the migration of design that is compelled by the collective psyche, in other words, culture. To think of them as the idea of a few designers, may not tell the picture of their significance for these similar applications are also inherent in culture. The reuse of motifs seems to be a cognitive phenomenon whereby desirable design character is recycled by the crafts.

---

[178] Lucie-Smith 1979.
[179] Not all furniture styles are architectural.

Supports in furniture can be represented with similar roles to a building's columns. Design analogies are achieved by various approaches, ranging from decorative carving to streamlining. This obviously obeys in part the cognitive criteria known as style, which is what makes associated pieces reflect each other. In Gothic design pointed arches are a device of the style (see Fig. 10) used in furniture and sculptural arts. The pointed arch and the church vault are echoed in furniture and detailing. The crafts may resort to carved discrete structural elements or just evoke subtly their forms, by suggestions through shaping, and this is still done today (as in Fig. 175).

From the observations of this study, it is clear that structures offer images that become a visual theme that is repeated much as a classic theme might be expressed in literature, theater or pictorial art. Given both the analytical character of structures and their association to walls, their images seem well suited still today for the appearance of many interiors. These may not always be meant to be luxurious, but just organized by some thematic classical criteria. Many examples of structural themes in interiors are observed in our restaurants, homes, offices, lobbies, etc.

An additional example that also deserves mentioning is the Baroque cresting, which is a fluid form that shows roots in a broken pediment and was morphed into a sinuous volute as those of Michelangelo's vestibule for the Laurentian Library; or before, as the volutes of Alberti's facades or the scrolls of Ionic capitals. The word volute is Latin and it refers to a shell. The Baroque cresting and pediment, as in cabinets and home entrances (and ultimately as in Italian churches), is a relatively large sculptural form that may give higher value to a doorway, or a piece of furniture such as a display cabinet and a bed headboard. It is today highly popular in East Coast residential housing and renovation, and it is repeated so much that it has the characteristics of a design fixation. It is more suggestive than a plain design, and to many users who want volutes on their doorways and display cabinets, it may reflect more workmanship than a rectilinear pediment. This cresting is perceived as an architectural form although it is not a structural type but a decorative one, and this architectural identity possibly contributes, together with quality wood, to make many of these pieces wanted for their aesthetic impact and eventually may even become family heirloom items. These meaningful old fashioned designs gain in value with age.

Structures cannot be separated from architecture, but many ornamental motifs display discrete structural elements by themselves, mostly of supportive type, such as columns, arches and pediment, but also roofing, thus we have made the point to refer to such images as structural motifs. While it is not straightforward why certain structural elements speak a cognitive architectural syntax so clearly in culture, two aspects in these elements may have something to do with it, deeply in the mind. One is possibly the sensory concept of support of a structure, such as its reliability or strength, and this is what a corner post informs. The other may be the association of columns and arches to the human physiognomy, which can be quite subtle and even hidden. In other words, some structural elements are subtly 'humanized' architectural forms (achieved through meaningful representation, which are not necessarily literal depictions of the human figure.) On the other hand, the importance of structures to life and culture is probably what leads to something figuratively meaningful. People become attached to prototype structures and their fame leads their image to meaning.

In the case of costume, the situation is intricate and is intimately related to the connotations of building types and clothes, in particular, ceremonial-oriented design. The route by which structure passes to this craft is not associated by trades that collaborate, but it arises in the operations of society. For example, the ancient goddess Cybele presided over the world's cities and she was depicted with a crown with fortifications or tower-like forms: the cities. Many ancient mythological sculptures in armor that guard a city, as Athena, are fortified allegories of the city. Cybele's control of the earth's cities is a strong precedent for the crowns of kings in ancient and post-Roman times[180]. The fortress towers depicted in later crowns have the connotations of supreme authority. The towers and city gates (prototype forms) therefore migrated straight from architectural meaning to head attire in the representation of Cybele, without the influence of technique or associated crafts. This is what meaning can do, and the Pantheon and the helmet are another case of form influence that is not dependent on methods or contact by trade. Thus, the forms of structures offer visual grounds of their own to encourage culture to make use of them.

---

[180] As ancient Roman emperors used a laurel wreath rather than a crown.

Figure 175: Bed headboard corner is streamlined to form a column and capital.

Contemporary design that evokes a column topped by a capital that grows out of the panel itself. Certainly a historicist (or traditional-like) piece in that it evokes classical style, and at the same time organic in that if flows continuously as a natural form would do. Design by Hudson Broyhill.

Chapter 14

The Place of Architectural Motifs in the Issue of Coating

In the nineteenth century, John Ruskin wrote about some practices he deplored in the treatment of stone in Gothic architecture, the use of iron, and he also referred to the cladding of metal-supported buildings with masonry as a falsehood. In this example, masonry is applied as a face appearance that masks the true support of a building. Ruskin tried to differentiate true and honest design from this non-genuine application of a face appearance that he sensed to be deceptive. To Ruskin, masking structure was a 'sin'. This example led to the development of a common notion that sees any coating of building parts as a problem in design. Possibly the eclectic and elaborate approaches to style of the nineteenth century gave license to produce some shams, as the age also blended historical styles at will, sometimes with highly wrought invented decorative applications without symbolism or sense, and this substituted a well-thought out aesthetics in all crafts.

Ruskin's opinion is valid too in terms of this practice being misleading to users. People who enter such buildings may think they are in masonry architecture, when they are actually experiencing an unknown support. And there are additional issues other than the proper approach to design, such as the right of the user to experience reliable and trustworthy architectural construction, irrespective of the greater strength of metal, or a user's levels of discernment of structural support in any space. Perhaps a subliminal perception of a large-scale artifact of heavy materials spanning above ourselves, makes the issue of confidence in genuine support, important. Ruskin tried to outline some exceptions to the coating practice in The Seven Lamps of Architecture, in art, however, Ruskin's criticism grew larger in twentieth century thought, and it ended up becoming all-or-none and being very influential. But in looking at the practice of building, we find a lot of masking of parts. Masking takes place in smaller ways in most construction. There are abundant instances in the covering of joints, in using veneers, or keeping wooden supports behind plaster, and so on; therefore some degrees of coating structure in construction must be more tolerable than others[181].

---

[181] A couple of examples coated in white stone are the Palazzos Cornaro and Grimani in Venice, by Sansovino and Sanmicheli, and the Pantheon's interior surfaced in marble and its exterior in bronze. Even painting and wallpaper are coats. Surely a purist modernist would reject these too, but we fall in this rigid attitude in the elimination of design resources, and there are problems for designers and creativity in denying them too many instruments.

Certainly this awareness of the problem of coating design can be seen as a step further in the development of understanding in design. It guided the values of good design in the contemporary age. However, everybody learned the issue as a universally valid code, sort of written in stone, and all design treatments were looked down upon, hence this view was not flexible enough.

Today, when energy is at a premium, and the earth is threatened by overbuilding, scarce natural materials are better off as veneers than in solid state. Additionally, a painting on a ceiling and a mural, are a coat too, and there has been painted illusion in history's great ceilings, and interesting perspective artifices in Baroque floor tiles, and in *trompe l'oeil* murals. In some of these cases the artwork of artists like the Tiepolo family's ceilings, is as valuable (or more) than some of the frameworks behind them. After the proposition that coating is wrong we have been unable to respond adequately to the benign examples of hiding some kind of structural parts behind other surfaces.

Design was not hindered by such limitations until the values of modernism developed from the nineteenth to the twentieth century. Neither a good writer nor a good theorist has to be one hundred percent infallible. We may remember the nineteenth century thought that declared that metal would never triumph in architecture because of its linear quality[182], for example, and we can see all around how incorrect this was. But the problem is not in something thought over a century ago, before the mind had a chance to figure out how to best work with the linear expression of metal, but it is in our faith in a design value of limited scopes.

Another coating practice needs to be acknowledged and discussed as it might be perceived as connected to this particular logic. A study of architectural motifs must reflect on equivalent intent and results between the coating of structure and the use of structural motifs as a surface coat in the crafts. See the reliquary in Fig. 1. The simple container is covered with an exquisitely detailed superstructure; hence it gives the box a different appearance than its original humbler origin. It may be additionally commented that architectural appearance and/or elements were certainly not created for these crafts, and strictly speaking, they do not belong in them. However, the type of work of this container approaches the level of fine art….. Its architectural identity is representational, and it is a precious object. For sure using building imagery as coating, by its nature as surface application, can be paralleled to coating in a building, but the development of motifs in culture is a phenomenon with very different intents from those of masking structure. One of the differences is that the craft is done for the needs of culture, and we can't emphasize enough how important the coating, which is representation, done for cultural purposes is to people, and therefore, it ought to be too, for the practice of design. The implementation of a face building appearance on a metal frame had a different gestation and purpose than the formation of structural motifs. It is a solution that avoided the effort –or the resulting design– of giving a metal-supported building a character of its own. Instead, the idea behind shaping a reliquary with the form of a church arises, not in avoidance to handle appropriately the entity's structure (as in a masonry-clad metal frame) but in a wish to make the artifact meaningful as a little portable sanctuary (that can be carried). A basic container gains immensely in artistic value by this treatment; it is a cultural treasure that is rare. And what is more important still, is that in the eyes of the faithful, the church-like box satisfies the sheltering role of housing for the relics that they want protected.

---

[182] Semper and Ruskin had this idea about metal.

In an object such as the miniature church of Fig 1, the building image makes a large difference in a simple container. It greatly enhances its expression and value. Yet an object as preciously-built as this one, is hardly known in the design field, as even though it uses a dome, a pediment and other structural elements, they are all part of the elaborate mentioned outer coating. A modernist may thus glance with curiosity and interest at this type of object at first, but his/her training may not permit him/her to see relevance to any 'correct' design in it, by noticing that it is a decorative surface. The untrained eye instead, possibly without awareness of the principle of not covering an authentic form, cognizes only a finished product. The amateur does not hesitate to enjoy the object as an artistic form. These different appreciations are in fact two ways of knowing, or grasping forms, and two different interpretations or appreciations of visual culture. Both are founded on reasonable rationales used by the profession and the public. One follows conventions established in the field which have not been inclusive of culture recently, the other, is a basic perceptual response. It doesn't matter to this work which way is better, but what matters here is to create awareness of these relationships which should be considered when judging the truthfulness, or the quality or merit of design.

The above case of fine craftsmanship of what is essentially a surface shows that some of our design values are not holistic. Decorative (or plain) design can be evaluated favorably as artwork, or unfavorably as a façade, according to the design position of an age, and the design values we believe in; but an object of this kind, aesthetically enameled, conveys more than a principle, as it informs that it is a rare and finely-crafted work of human hands. A coating practice, or even masking, may have other intents than to mislead, and this may be extrapolated to architectural examples we are judging –as we have mentioned in the case of doors, balconies, railings and chairs that don't function. In the craft, coating is done to depict an image, and not a functional construction. The container of relics does not house human life, but it only keeps its remains. The housing (storing) role of a miniature building seems appropriate and reverent for the head of the saint it once housed[183]. The creation of this miniature was compelled by respect for one of those individuals that the living could not bury and could not let go. On the other hand, a design practice that leads to misinformation about a building's true support is a poor solution that should not spread in the profession, as it can lead to other *laissez faire* in design. But the container that looks like a church cannot be deceptive in the way a functional edifice is. In fact, the two reliquaries that were made in this form in history, recorded the aesthetics of the umbrella dome. This dome has no extant equivalent in a building in the West with the integrity of this architectural miniature. Umbrella domes have been remodeled in later times and their original structures exist in fractioned conditions and drawings. The architectural motif in the reliquary recorded the exterior of this attractive early dome that was obviously admired in its times, and therefore, the world can see it – an important unspoken purpose of the migration of design is to prevent the loss of a treasured form. Though this dome type was never intended for this end, its reproduction in a miniature was good, as it ended up as an indirect way in which anyone can enjoy this dome.

---

[183] Saint Gregory of Nazianz.

Photo by Chris Kausel

Chapter 15

The Suggestion of Precious Materials

Materials give the best results in particular applications. For example, today in the midst of industrial manufacture, wood can still lend itself to crafted looks, something that is observable in both traditional and contemporary design. Aluminum and plastics are certainly modern materials and are used in fabrication processes where both the material and the process determine the looks of products. While there are some aesthetic minimalist aluminum and plastic examples, such as some graceful designer bar stools that can come to mind, the perception of these materials is usually not comparable to the perception of natural materials. However, the new industry of trans-materials may achieve a type of industrial fabrication that may even surpass that of natural materials. Examples might be the organic designs that can show transparency, colors and extra hardness and durability, in quartz, cast stones and polished granites.

The finest materials have been used in the past in some statuary and ornaments intended to be just objects of decoration. These can be sculpted in alabaster, porcelain, glass, fine woods, silver or other metals, and may be used to evoke sensory recollections for a fine visual effect in an interior. The Asian and the European traditions offer porcelain vases and figurines with animals, flowers and musical instruments. Some Rococo figurine types have been reproduced too much and may not appeal to many contemporary people, but may nevertheless be appreciated by some buyers as a rarity or antique. The modernist versions of decorative sculpture can be abstract and free-formed. It is common to see fragile and very delicate materials in these objects that have no other function but to decorate. This is part of an intended impractical luxury. Visibly breakable and rare objects seem to capture attention. A viewer feels compelled to glance at something that is both attractive and extremely fragile. The fragility of these statuettes and other similar objects probably quickens appeal for these crafts. There are notions of fine kinds in crystal for example; and not only because it is fragile, but also because it is transparent, and this connotes qualities too; that include the beauty of clear water, evoked in the solidified material. Transparency also can recall some life or living parts in some applications, such as the eyes, insect wings and translucent water microorganisms. These recollections can be mostly never thought about by anyone who is born accustomed to glass, but the aesthetics of glass is related at least, to the recollection of ice and water. An important feature of glass is that it permits to see through it, and from here, it allows the control of flawless cleanliness in what is made of it as well. This detail can be highly desirable in table pleasure. Clean glass is agreeable to the psyche and vessels of clear glass are good for drinking pristinely-clean liquids, including water, as our mind is weighed down by the knowledge that even fresh water is becoming scarce and polluted with prescription drugs. Some utilitarian cups are designed within the limits of fragility and very slender glass stems, letting us see their desirable thinness. Glass cups refine the table pleasure offered to guests by their sound and looks; hence, they enhance our social attentions to others.

Fine materials boost the appearance of a space where they are properly displayed. Objects made in the finest materials display their sensory qualities that translate into refinement, wherever they are placed. Museums collect these objects too (See Fig. 176). If a figure of this kind reaches centuries of age, its value as rare object increases.

To the above, we must add that the aspirations of craftsmen also can influence what these fine materials connote. The wishes, inspirations and intents of craftsmen can be expressed in their creative work sometimes, and though many go unnoticed, some may be perceivable qualities of the art. Fine materials are first class resources for suggestive qualities, and craftsmen can use them to create delight in design. A few illustrative examples should suffice; a craftsperson may wish to give to a vase the status of a royal piece, or a matte softness that evokes skin, or attributes that have an appeal of something antiqued, or delicate, and so on. Many of these evocations are perceived by users, or at least will feel attracted to such effects and see in them agreeable sensory qualities that look nice in a space. Such finesse reveals at times what should be excellent attributes for the interiors of buildings. When structural forms are sculpted into precious objects, the adapted architecture as decorative motif is more polished than in an actual building prototype. We can notice for example, that a reliquary is made of valuable metals and ivory (see Fig. 1), whereas a building prototype –as we know it today– cannot afford such expensive materials because of its scale[184] and exposure to weather. In the past towers in miniature were carved in ivory, today censured, and in the nineteenth century there was a dream of crystal palaces, as shown by Paxton's exhibition hall, and the twentieth century utopian glass city. A Wedgewood vase's columns are much softer than stone columns too. An architectural mahogany cabinet is more finely polished and varnished than any stone façade or even an interior wall; likewise, a porcelain object bearing structural motifs has a softness that is unachievable in concrete or even plaster. While many of these finishes are not utilitarian, they nevertheless compare to the case of an imaginary architectural pavilion of medieval iconography, in that they seem to sometimes inform desirable and imaginary conditions for an idealized architecture that belongs inside. In the special effects of decorative ornaments, we find cultural feedback about desirable and even ideal architectural features. The garden pavilion in white porcelain of Fig. 176 is a structure intended to be as extravagant and open as possible, and utility is not important in its case. Only its organic soft suggestion of a search of the mind counts. The intuition of the Rococo age came up with decorations of this type. At the Meissen porcelain factory where it is displayed, it is executed in a large scale. Considering the fragility of porcelain for its scale, we can see that it properly represents an extreme form that an age obsessed with design produced, possibly for the user who delighted in rare pieces.

---

[184] There are ancient texts that describe silver and gold in cities and buildings though.

The mind conceives essential elements, such as columns and arches, and these in turn may take the direction of imaginary supportive concepts in time. We may see in the iconography of these significant structures –as in some rare objects– how the psyche expresses idyllic design in soft materials. In fact, some man-generated materials, as porcelain, whose earliest known origin was China (it took a long time to Europe to come out with a similar type of paste), may have been thought about after exposure to the beauty of natural soft materials such as the 'mother of pearl' (see Figs 177 a to c). The interiors of Rome were lined with a soft material, marble, and the European and Islamic interiors are conceived in smooth stone, plaster, marble, tile or *escayola*, that faintly and subliminally evoke sometimes membranes, nerves and wet surfaces, whereas exteriors can be plain and sometimes even left untreated (see Figs. 45 to 51; 69 to 71, and Figs. 119 to 120). A shell is usually rough on its exterior and smooth interiorly. This is the action of collective intuition. The solutions of the mind for architecture unwittingly mirror those of nature sometimes.

For all the reasons observed in this study, whether some of us may be able to discern these relationships or not, it is not fruitful to our knowledge to discourage the practices of embellishment, lest we want the domain of design influenced by the internal psyche to remain buried.

Figure 176: Extravagant pavilion of garden type.

Executed in fine porcelain. Rococo Inspiration. Meissen Factory display.

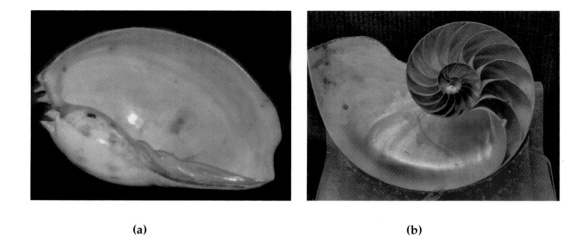

(a)   (b)

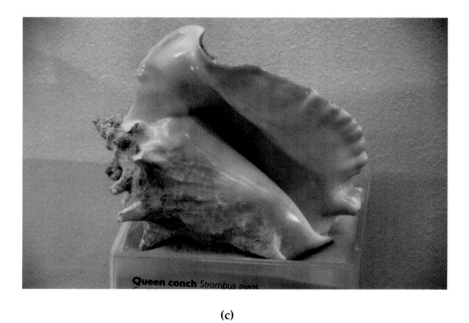

(c)

Figure 177, a, b and c: A shell's soft material.

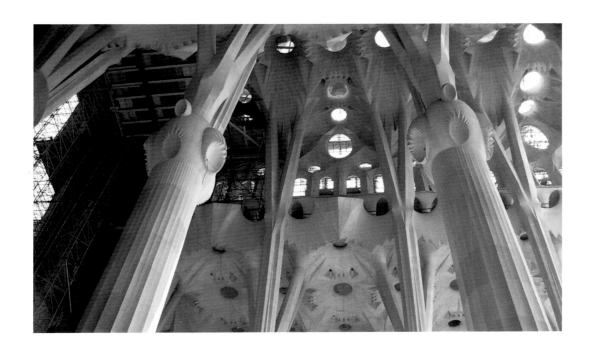

Part V:

DESIGN ISSUES OF TODAY
AND YESTERDAY

Foreword to Part V: Why Culture Decorates

The activities we call decoration show an inherent inescapability that makes them an aspect of social culture. By this we mean to say that decoration is a habit of culture that is favorable to social relationships. The psyche is social and learns mannerisms and dexterities to suggest agreeable atmospheres and show consideration to others. People improve and decorate as a means of delighting acquaintances through the presentation of their reception areas and gifts. A saying tells that every gift is a wish for the happiness of the recipient. The individual who embellishes often makes an effort to offer something above the ordinary to others. Thus, a streamlined glass table or a crescent-shaped bread roll, are both subject to a figurative process of transformation that makes the presentation of the product appealing and nicely-presented to a target audience. Some individuals decorate at a professional and artist level, and some don't bother, and focus on other tasks they carry out in society; however the sense to embellish for propriety is present in most normal persons. Persons in sound physical and mental health tend to beautify their spaces, and not in vain the crafts are used as a therapy for emotional and other problems. Improving personal spaces is a habit, as cleanliness is a habit, and it may include activities such as the practice of crafts, adding or removing objects, coating, grooming, or displaying flowers. This many times translates into urbanity and the primary targets of decorative efforts are people's personal appearance and the presentation of living and working spaces.

There is no question that cutting down in excessive decoration (as that of the Victorian age) was needed as the modern society of workers took over the agrarian type. However, since changes in style transform the appearance of the world, we grow accustomed to contemporary images and our perception of earlier styles can become distorted by our focus on current looks. Too much decoration must have been encumbering to comfort, this is true. However, if we are inclined to believe that the embellishing custom is unnecessary in design, and so we teach our students, we are left with many unresolved and perplexing issues: such as how to account for millennia of unnecessary embellishment? How to explain enhancement even in modernist forms? Why do we dissociate in creativity the finishing activities from construction? How did decoration and imagery develop and what for?

Decoration is certainly a practice that has accompanied human culture for the length of its existence and must have arisen in the mind's industriousness, being created along with key developments, such as the inscriptions used in early tablets and the propitiation of the dead with flowers and shells, for protection. Incising marks on objects, and adding fanciful forms to head attire and armor, helped early culture to distinguish ownership, leadership and pecking orders, and to summon spiritual help through magical symbols. However, the decorative habit was bound to develop a compelling purpose, and this is the design and enhancement of appearance. Practicalities aside, everybody knows from personal grooming that the improvement of appearance is a necessary routine so as to be acceptable to others. This becomes involved in the important social behavior by which each of us is part of a group that has a minimum of expectations of its members; for example, a working environment[185]. The non-practical treatments in architectural design also transform plain or ordinary appearances into something presentable and sophisticated to others.

---

[185] Sometimes this enhancement of appearance is directed to enhancing physical beauty and youth, but not always, as other aspects of appearance may be sought for other reasons, such as appreciation of vernacular forms, clothes patches, or aged fashions.

As a design 'truth' dissolves in time it may be contradicted by another that arises to take its place. In the twentieth century, design was said to be made 'pretty' by adding decoration to it, with the connotation that prettiness was no longer a characteristic of a sophisticated taste. Embellishment is colloquially identified sometimes as feminine in its character. The references 'cosmetic' and 'dressing' that are used in renovating and adorning a building for festivities, may enhance the perception of surface treatment as feminine[186]. These associations reflect that styles are always an outlet for the social dynamics of pecking orders and biases such as the need to feel superior to others. A certain taste tends to be used as an instrument to stand out; and one of the ways to stand out is to look down on the characteristics of the taste that lags behind, or that of other castes. Thus, design syntaxes, like many other characteristics of culture, fall prey of social stereotyping.

The social dynamics that develop around decoration also affect abstract design. For example, design-savvy individuals may have stayed away from decorative design in the recent decades, but if we lived by a decorative spell, they would likely want to decorate. Under a decorative spell, design becomes prey of redundant embellishment, and under an austere spell it falls prey of the lack of it. Social pressure affects the values of taste, however social dynamics are not an aspect of the sense of vision. Discerning this point has been important in this work. In spite of the force of social circumstances, design –and in particular, its merit– is not exactly the product of our social behavior, no more nor less than the quality of cuisine is the work of our hunger drive. Fashions change our taste but not the long-term taste of all mankind in history. Just as in any drive, social behavior establishes wants and needs, but it cannot create outcomes. The social posturing of the complex world of design can blur the understanding of what is good to a majority, and this keeps the door to the knowledge of culture and creativity through design shut tight. Regardless of design values and positions that avoid embellishment in architecture, decoration, though time-consuming and encumbering as it is, has been produced in the crafts during the staunchest bans on embellishment. Hence, despite the implementation of values to depart from it, thus far, this has not happened.

The embellishing habit may be a natural inclination, and as such it needs to be known and understood for what it is, and what it accomplishes in visual culture. The profession of design should increase its knowledge depth by not falling into the trap of closing a door to everything associated to common taste. Furthermore, the studying of vernacular visual culture needs not mean to implement it.

---

[185] Much of what seemed too pretty to the early twentieth century was refined to the Baroque and Rococo ages, when people were compelled to place greater emphasis on decorative appearances than we do. French taste dominated during those periods that were clearly an age of design and of people obsessed by it. Fashion can be an explicit type of design to illustrate this point. Baroque gentlemen (including musicians and men of science) wore long-haired wigs with curls, high heels, lace, and colorful silk overcoats that enhanced the waist line. The official paintings of Louis XIV, for example, show him with red shoes with ribbons, and French generals were painted with a large ribbon with a bow on the hips, and this extended to the 19th century judging from the 19th century portraits of Napoleon as a fighting soldier.

Photo by Chris Kausel

Chapter 16

The Authority of an Ancient Structure

We have introduced the authority of the mind on design, exemplified in the structural and decorative accomplishments of Gothic architecture. In this section we shall analyze the authority of some ancient forms, which are possibly in some intimate relationship with the mind. It can be observed that design forms keep on living in culture even long after original trade practices were left aside. This phenomenon is observable in the contemporary industrial manufacture of historicist building parts in doorways, windows and interior elements. As technology becomes sophisticated, it has freed popular designs from original methods and parts are mass produced in period styles. This historicist effect of modern manufacture was not planned but it happened from demand for images that the public prefers, and it has been criticized in the field, nearly as much as the early Victorian manufacture of antiques has been.

Those who believe in the superiority of modern design have been waiting for the larger public to enjoy functional visual paradigms, however the still thriving traditional architecture and interiors indicate that many people continue to feel inexorably drawn to the homes and furnishings of their forebears and history; or to beauty-treated 'modern' models (today part of design history), such as those observed in Scandinavian design, or the likable iconic examples of the twentieth century. The design interests of laypersons have been variable during the last nine decades, trying to embrace the sometimes abstract vocabularies of the twentieth century and sometimes hesitating or changing their minds and sticking to traditional and mostly regional forms. Interest in both regional design and curiosity for novelties tend to be effortless in the public.

Concentrating in the behavior of the public to traditional design, we can observe certain trends in the market of home goods which are also observable in architectural design that for sure depends on manufacture of parts such as windows, doors, stairs, water fixtures, flooring, roofing, etc. While the market of traditional building parts does not directly impinge on the output of a design office, the contemporary market mentality permeates much of life and indirectly may affect professional design.

A market needs renewal when consumer interest diminishes[187] as this ebbing of consumer enthusiasm causes the sales of certain products to plummet down. Weakened consumer interest is thus reawakened by frequently making available novel supplies, and by large inventories that provide even a higher number of alternate options than what is shown on a floor and can be ordered from catalogs. The industry seeks to maintain the momentum of the novelty, in particular, by habitually offering new products. If no true innovations appear in a year or two, many new products are still offered in the market that consist of pre-existing items with minor revisions, as change (any visible turnover of products) is believed in the world of business to be an indication of a thriving industry; and the wavering convention of what's in vogue is largely subject to these dynamics, and advertisement. As new materials create niches for revised products the market turnover of goods increases, and this has certainly been the case with the high-performing building products that have stimulated retrofitting of existing buildings, especially for saving energy and water. The commercial explosion of digital products has been another incentive to actively add computerized equipment, and our new century's digital-looks are attractive and convenient to the public.

The contemporary design values advocate differences from the design of history that affect a majority of modern products in some key features, for example in the simplicity in the general lines of design. This may change without notice, but for the last ninety years simplicity has reigned supreme. For one thing, it offers advantages to both users and manufacturers, it can be less labor-intensive than an elaborate appearance; and for another, smooth ergonomic edges in personal furnishings don't grab clothes or cause bruises and are easier to clean than highly textured pieces. New design can afford to be not only simple but anything designers want it to be, as the sales of design are now aided by stunning presentations, and by aesthetic lighting, photography and skilled composition of pieces, so that shopping windows and showrooms can sometimes display simple boxes or crude concrete blocks, or recycled scrap, all in attractive presentations. New approaches to fashion elicit interest in buyers, even if it may amount to only a new application of old and even worn-looking items, or those that were not thought to be attractive before, but whose special aesthetics or 'cool' looks become newly-discovered. Our spaces need some variation to make us feel renewed and satisfied with our surroundings, so our age shops for new items and refurbishes places. But in the habitual keeping of the pulchritude of our environments, we also stumble upon the old and the mysterious imports of timeless images, and re-discover them. Novelties, exciting and all, have not caused new generations to leave behind the unoriginal, solemn, enigmatic or romantic traditional forms. Abstract, plain, streamlined, ergonomic, high-tech, or otherwise advanced forms, as attractive as many are, have not caused society to leave behind the looks of historicist pieces yet, especially in dining rooms, ball rooms, hotel lobbies and other fine interiors.

Society is today choosy, spoiled by market novelties, but in spite of this, the same old traditional designs still live in our culture. In some ways then, the response to old images seems to be a different phenomenon altogether from the dynamics of novelty and the promotions that accompanies novel design. The success of old styles, i.e., the 'classics' and iconic designs, have been explained variously, not only as a kind of nostalgia, as discussed earlier, but also as something that is tied to social pressures, such as a search for prestige products, or people doing as others do, a wish to stand out from the crowd in refinement, and other typical social-based

---

[187] A drop of sales can be counteracted by a shopping area renovation. Recently, malls are seeing complete architectural renovation just a few years after they were constructed.

The Permanence of the Arch

(a)   (b)

Figures 178 a and b: The Arch –Then and now. *Source: Author's undergraduate course Building Construction and Materials.*

Wooden models of arches. Voussoirs are not glued just as masonry arches stand without mortar.

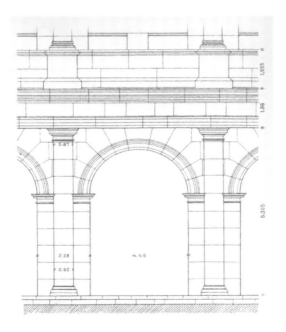

Figure 179: The monumental order of ancient Rome in a stone masonry arch flanked by columns.

Elevation of Roman arch in the Colosseum of Rome. Von Egle.

Figure 180: Contemporary arched windows manufactured with modern means for common homes.

Figure 181 and 182: The arch's multiple glass forms in mass-produced windows and fanlights.

Figure 183: Masonry and wooden arch in a common historic entrance.

Nineteenth century neo Renaissance door. *Colonnaden* St. number 5, Hamburg.

Figure 184: The symbolic outline of the rising sun in a fanlight above a door.

10 Downing Street, London. Almost three hundred years old, this building was first Hampden House and later Kynvet House, until it became the residence of Prime Minister Walpole. Author's enhanced diagram.

Figure 185: The radiating fanlight in vernacular version.

Small original vernacular door at Ely England.

Figure 186: Radiating ornamental pattern comparable to the fanlight of ancient Mycenean origin.

Treasure House, second millennium BC. V. Egle.

Chapter 17

The Public's Protectionist Attitudes

Quite evidently we no longer have opportunity to scrutinize the opinions of those who created the structural motifs of history. However, we can infer that in order for structural motifs to be created, structures must have been substantially aesthetic or otherwise exemplary for carpenters to reproduce their visual elements. In many ways, the continuity and reuse of design presented thus far is also expressed in the general wish of laypersons to preserve monuments. Conscientious preservation is a relatively recent idea that was launched in the nineteenth century by the need to restore the medieval architectural heritage of Europe. Early conscientious restorers were Eugene Emannuel Viollet-le-Duc and Augustus Welby Pugin. Before the professional activities of these scholars, the interest of Raphael in the archaeology of ancient Rome, in the Renaissance, the eighteenth century discovery of Pompeii and Herculaneum; and the archaeological activities of the early nineteenth century; have also been important in a global development of attention to valuable original structures.

With regard to this topic there are historic activities that reflect indirectly and sometimes directly the wish of earlier people to care for the buildings they they didn't want to lose, and this wish must have been manifested in world regions at any time. Whether this was in the past part of public action or just the idea of few individuals, we can only be sure of the latter one. Public opinion counted very little in history. This fact can be deduced from the forms of government and social conditions that predominated in history. To the many people to whom life was hard, the activities of protecting buildings was beyond their immediate priorities. Few individuals must have decided the fate of buildings; and they were primarily those in authority. However the action of those in power may be grouped as the action of laypersons and this detail is important to our focus on the attachment of amateur individuals to traditional design.

Indeed, many priceless buildings that have survived all the way to our current century could have been lost after a generation's change of opinion. Buildings were destroyed in warfare and social revolutions and could have been lost for good, as the removal of a government, or a faith, brought much burning, destruction and hostile feelings against the styles of regimes deposed.

However, decades after a revolution, as economies took off and communities wished to beautify their environments, there were those who wanted to restore the original splendor of the best damaged buildings, including those of a regime displaced. Decision-makers have admired significant architecture and decided to keep a record of their development and construction. Committees have deliberated and decided to keep track of the know-how of edifices and their crafts. Some lost buildings were completely rebuilt. An example is the Williamsburg Capitol reconstructed in 1927, and a recent example is the *Frauenkirche* of Dresden. There are other cases around the world. The recognition of design merit, aesthetics, significance, fame, are important movers in culture that lead people to maintain architecture in the environment, of not letting them go or become neglected or displaced.

As a result of the above public action, neither the obsolescence of styles nor arising creeds have caused societies to lose track of many of the exemplary structures of history, although evidently, many buildings have been lost, in particular, in the early times of the first millennium, but in the long run, changes in mentality have not yet kept the culture of old design from being continued. All of this gives a picture of a cultural habit that has had great persistence in the midst of great challenges, since those who should know design best, the design experts, often notice a backward-looking dimension in the focus on past forms, and thus, deplore that society's vision dwells on what has been created rather than on what is to be created. There are surely other components that are not related to design in the public's protectionist behavior that must be taken into account too, such as consideration of the age of old things, and their value as historical records that are still extant and need preservation, and in these interests there is wide agreement. Thus, both cognitive and learned behavior persuades culture to save their cherished structures whenever wars, new ideologies and even stylistic obsolescence threaten their integrity.

Detail of a drawing by Schinkel

Chapter 18

Invention, Inspiration and Biomorphic Design

We have already briefly mentioned that there are values which establish that the form of architecture is best when it reflects only methods and raw materials. This is of course an ideal view that largely represents a design outcome based on the solving of forces and requirements. The post-and-lintel system and the true arch, for example, are basic construction systems and they have respectively a distinctive right angled and circular form. They were created as a framework and for spanning large spaces, and their basic shapes represent invention untainted by embellishment, if wished. But imagination will use other existing resources at hand too. Apart from using a design concept as project guidance –that resorts to many sources of inspiration and is taught in the studio— architecture has resorted to biomorphic designs and other existing built concepts to create structures. A biological form can suggest an idea to an architect that can be tested as a supportive system. Examples of biomorphic design in architecture abound in the work of Antoni Gaudi for instance. His imaginative designs blend natural forms with his 'equilibrated structures' such as his catenary vaults. His organic forms have been interpreted as trees in his *Sagrada Familia* church (see Fig. 187). It is true, though that the genius of Gaudi remained unsung for decades before it was recognized outside of Spain, and this was to a large extent one of the twists caused by the focus on functional and abstract design of the last nine decades. Gaudi's design was also seen over such decades as intentionally bizarre, under the pressure from his patrons to bring attention to the Catalan region's separatism.

Then, the form of a scallop is another natural form that has been incorporated abundantly in architectural niches since ancient Roman times. Since then, Western cultures have depicted niches shaped as a scallop in architecture, especially in interiors, and many of them are reflected in Renaissance painting (see Figs. 188 to 194) and even in chairs (see Fig. 8) and fountains. Few people today wonder why this motif has been repeated so much, and rarely would anyone think that the scallop might be significant in architecture. It is perceived today as plain decorative. Because it was more intensely represented in the Baroque, many see it as an aspect of its stylistic syntax that re-inspired the shells designed in the Victorian age. However scallops were present in all ages of classicism and this includes ornamental depictions of ancient Greece. The 'frivolous' scallop could not be inconsequential to architecture to be so consistently depicted. It likely provided an early model idea for a concave space, to experiment with, judging that the scalloped niche was built in Rome in the size of exedras, heralding the gored domes[193] (see Figs. 195 to 197). The dome proper, of vast scale and derived from the semicircular arch was first built in Roman construction and gained popularity in Byzantine times. Cupolas were a common architectural form of Islam, and were transformed into gored domes. The Great Mosque of Kairawan, Tunisia, has the earliest monumental example of a gored dome. The segments of the gored dome are concave on the interior, as if scooped out. There are some small versions of a dome of convex rather than concave segments, as in the Minaret of Kutubiya, Marrakesh, which are referred to as a melon dome. There is a gored dome in the *Sala de La Rauda* at the Alhambra that was designed with uncommon ribs of an organic appearance that flows to the concave segments as those of a conch shell (see Fig, 195). Other domes of Hispano-Mauresque manufacture in the Cordoba Mosque appear as if cast after a mold that may evoke fruits and flowers as well as shells (see the central and smaller cupolas of Figs. 196 and 197). These domes or cupolas are made of brick masonry, and their presentation is related to a clay mold. If a scallop is found on wet sand at a beach and picked up, it leaves a mold of its form in the sand. This is a common event that may be part of memory and could have brought to mind the investigation of a shell form as a structure to these ancient builders.

In any case, where a designer experimented with a natural form, it was necessary to translate it into a static system. These explorations into structural possibilities show merit not only in translating a form into a large scale structure but also as investigations of the supportive properties of certain natural shapes, that not only are far more sophisticated than a post-and-lintel assembly, but also can prove to be efficient in a proper structural application.

Finally, there are instances of construction inspired by other constructed structures in culture, such as a ship's hull that was adapted for oblong vaults (see Fig. 198), or the idea of a bell shape in the dome of the *Frauenkirche* of Dresden, and even the reapplication of a civil structure in a different context could be cited here, such as an aqueduct's arcades to support a building on the interior. Hence, inspired structures have not been uncommon. Even some modern architecture has been –at least visibly– inspired by other forms, such as the modernist chimneys of boats, and the general appearance of the early jukebox[194].

---

[193] Hersey, 1937.
[194] In the streamlined *moderne* and Art Deco examples.

Inspiration in Natural Forms in Design

Built after Gaudi's last model ca. 1925.
  Perceived as tree-like piers, organically tilted, vaults and wall surfaces were made as hyperbolic parabolloids. Star or flower patterns form the ceiling.

Photo by Paulina Kausel

Figure 187: Antoni Gaudi's nave of *Sagrada Familia* cathedral.

Figures 188, 189, 190: Shell-formed niches

Ancient shell niches were used in hemycicles at pleasure gardens and in tombs.

Left: Illustration from a Roman inscription , in A. Rich, Dictionary of Roman and Greek Antiquities, 1893.

Center: Shell in niche of human scale at Montserrat Monastery in Catalonia, Spain (boy shown is eleven years of age).
  This Benedictine monastery was probably built in the 12th century. It was destroyed during Napoleonic warfare and rebuilt in the 19th C.

Far right: Chalk depiction by author of Sansovino's Apollo, Logetta of Florence, 1537–45. After photo in Britannica Encyclopedia.

Figure 191: Renaissance and Baroque paintings feature many niches with a shell.

Madonna and Child. Domenico Di Bartoloca, 1433. Among the important symbols of a shell are origins, protection and rebirth, as in resurrection after dying.

Figure 192: Shell niches with the shell form in different directions.

Detail of the central panel of Saint Lucy altarpiece by Domenico Veneziano. The Virgin and Child with SS Francis, John the Baptist, Zenobius. Uffizi, Florence, c. 1440.

Figure 193: Shell Niches.

Novitiate Altarpiece with Saints Cosmas and Daman by Filippo Lippi, 1617 to 1750.

Figure 194: The shell niche in nineteenth century architecture.

Schwerin Palace. Germany, nineteenth century.

Figure 195: Gored dome at the Alhambra, Granada.

The sixteen ribs of this dome flow smoothly into the gored segments and the bricks were cast in curving form so as to grow gradually toward the perimeter, as in a natural form. The more common dome ribs are arranged as bands of rectangular section, unlike the skin-like ribs of this dome, that must be considered an organic approach to the assembly of a structure. The idea of these ribs may come from observation of a scallop shell's 'ribs' that flow smoothly to the gored segments.

Figure 196: Gored dome evokes the geometry of a flower and some fruits.
Source: Wikimedia "Cordoba_mosche_innen5_dome"

Figures 197 (a and b): Example of traditional ceramic form in a clay mold and the small cupolas and squinches of the Cordoba Mosque. Right: 'J. EntrenasIslamic cupola'

Inspiration in Other Constructed Forms in Design

Figure 198: Ship hulls have been proposed as model for a type of vaults of Spain. *Sala de la Barca*, Comares Palace, Alhambra, Fourteenth century.

> This vault is oblong and is said in Granada to be based on a ship's hull shape, and from this technique comes its name Barca.
> Framing of the Sala de la Barca. *Drawing by author*.

Chapter 19

The Illusory Dimension of Categories

Cultural interest in some exceptional types of design is detectable, however, the relationships it involves are not easy to formulate. A major difficulty in trying to observe some converging public proclivities is that wherever there is agreement in visual preferences, there are different explanations for the choice. These different views are for the most, semantic in nature and arise from imprecise understanding or explaining of the material reality of design by the multiple interpretations of different people, and from different points of view.

Labeling design in necessary categories in itself is the result of the functional features of verbal (or linguistic) communication which attempts to reach clarity by grouping ideas sometimes into opposites. A description may be objectively done however the act of verbalizing a category's characteristics may obey many aspects of culture that are not material, and thus not be accurate to some visual effects observed. Conventions are used all the time in interpreting design, and many conventions are idealistic and not necessarily germane to the sensory impression of design. In our linguistic-conditioned understanding, historic and contemporary looks, for example, are usually perceived as opposites of each other. This vision seems to originate in the fact that modernism tried to secede from the styles of history in about the '1920s, and in doing so it strongly rejected connection with any past style. Hence, in ninety years we have learned to make the same distinction. But the cognitive and sensory qualities of creative modern design, that needed to break away from the model, may not have followed all the expectations of its school and some examples even offer sculptural soft curves, as historical design once did. Categories help us learn and teach but if they do not reflect material circumstances they can cause discrepancies between our deeper seeing of artistic forms and our understanding of them in words. Differences between styles certainly exist, but styles are not always executed with opposing design features (and why should they be so). Yet common understandings of design maintain a perception of opposite values and sometimes even rigid positions, and thus can lead to stumbling blocks in description.

This next example illustrates how categorization creates beliefs about design that don't necessarily have substance. We can introduce a case of a typical conviction in an individual who may say that s/he has a clear penchant for just one single style, and let our example be finely-polished and curvy Baroque furniture. This individual doesn't bother to explore other forms because it seems futile to try to like something else. However, the same individual may be surprised in time to fall one day for an organically-formed seamless countertop just out in the market that for example, is dark and polished so that its surface is highly reflective. This individual loves this twenty-first century design immediately and feels puzzled by this event. A contemporary type of design once thought incapable by its principles of moving his or her interest, can nevertheless be executed in smooth and sinuous approaches that change the previous circumstances. The puzzlement of this person derives from the bounds of learned conviction about the characteristics of contemporary design, namely utility, functional methods, little treatment, etc. Both, Baroque and organic design offer elements evoking bio-mimetic forms to the mind, and this is an important detail, as they may present something that could be attractive to the deeper psyche. The new glossy, organic, seamless forms of our current times can be masterful in their sensory dimension. Polish and shine are clearly design treatments to begin with, and of a sensory type. They belong in the world of embellishment and not in the modern search for industrial forms and untreated and crude materials. Hence, the seamless sleek organic countertop offers some of the characteristics of history's styles. In any period of design, the best examples produced may transgress its idioms, as talented minds may need to do so in order to express what they envision.

The above event is important in a cognitive way and the effect of the above problems can be made even clearer by translating this case into yet another example, a Baroque image with even more life-like energy than the Baroque cabinet. Bernini's bronze baldachin, though not the earliest Baroque achievement, reaches a type of form that defined Baroque architecture. The columns of this baldachin show a heavy muscular snake-like movement which has a dynamic twisting that evokes life (see Fig. 67). This meaning is present in the form itself, and it is part of a different process than that of the search for the columns of the biblical temple of Jerusalem. A previous section of this study introduced the idea that supports are an active medium that sometimes suggests its operation by resorting to ideas of vitality and energy. Bernini's columns are a perfect example of design vitality. The offspring Berninian designs that followed these columns in Spain, used black marble and polished the material to a 'wet' shine (see Fig. 68) and thus made the same type of forms look like umbilical cords that also seem alive (see also Figs. 69 and 70). Umbilical cords are a type of meaning that is also present in the twisting forms themselves, regardless of an absence of recognition of this iconography. These cords are certainly a very ultimate form as a symbol of linkage (in greater meanings such as in generations, continuity of life, heaven and earth, or ages of the world). If today, almost four centuries later, a glossy organic stone, quarz, or glass top may too bring to the subconscious the impression of a living creature, such as a shiny water animal for instance, it can surely reach the mind of an individual who appreciates the life-like forms of the sensory Baroque.

The above account hopefully illustrates the different understandings that sometimes can exist between verbal information and cognitive-sensory perception. Even though under clear linguistic categories the Baroque and today's polished and flowing countertop are separated by centuries, both, organic and Baroque forms have an important common characteristic however: their life-suggestive vision of design. Sometimes an unrealized evocation in some design captures our attention, whether it has been achieved by a material and its treatment, or by a given form, or it is a connotation of the entire composition. The forms of history have been successful in part, because many designs offer something important to the psyche, even if those evocations may not come through consciously.

Important cognitive and sensory equivalencies make design free rather than bound to conventions. This is so even if styles are tangible phenomena that are correlated to periods. Any organic form of today is more comparable to a historicist piece of simpler lines, such as one of Art Nouveau conception, than to a right-angled early modernist one.

Chapter 20

Objective Study of Form

The methodology followed in this work has been the observation of the cognitive roots of design, focusing on the generative influences of visual relationships and on image migration and transformation. The visual and sensory domains of design operate all the time, and are inherent in the mind and in human behavior. In pursuing our cognitive approach to the study of design, it is necessary to sort out what are useful means to assess its objectives.

If design is judged by our non-cognitive systems of reasoning it may be very difficult to elucidate much about the aspects of design which are founded on visual and sensory relationships. A lack of study of these relationships, may lead everybody to categorize design as a function of ideologies, and the responses of people to design, as a sole function of educational level and the elusive and occasionally arbitrary understanding of taste[195]. The road to the study of the proclivities shown by the public is additionally paved with obstacles arising from beliefs. How these could interfere with a cognitive approach must be carefully discerned and not be allowed to give a different interpretation to the suggestive magnetism of design. This type of inquiry must be concerned with what grabs the interest of any and all people, and is observed from the types of design that are acquired and admired in spontaneous types of reactions. In front of the multiple possible routes available to interpret design we must guide analysis to the domain of visual impression and suggestion. To achieve this we must first acknowledge what type of opinions will lead us to ideological snags that obscure the visual understanding of design.

1. First point. Some writers have expressed that replication in design is imitation, and this leads to the conclusion that there is no imagination in design emulation. This also acts as a dead end to investigation, for example, of the dynamics of migration in design. The understanding of this phenomenon lies in the study of the very relationships that have been categorized as imitation. Form migration certainly includes mimicry, and in no way we mean to deny this fact. It has been

---

[195] As the world's culture has increased and become a global village, the understandings of good taste have multiplied and are highly varied.

mentioned in the furniture design literature that the needs of buildings for identical parts makes allowances for imitation in design. Correspondingly, a situation in the crafts, where cabinetmaking skills thrive with good copyists –and where innovative designers might even be selected against– has been mentioned too, both as being behind the repetition of design and the styles of history. However, as we have stated in the Introduction of this study, the perpetuation of certain designs is ultimately compelled by the success of design among buyers. Form migration is also connected to inspiration and emulation of design, which have sometimes also been considered imitation. But inspiration and emulation should be observed to be more than design repetition. We must see its symptomatic character too (or any other significant cause of the phenomenon). Before industrialization the creation of equal pieces was done by craftsmen themselves, this meaning that design reproduction is not only the consequence of equipment that creates only identical products. The practices of trades themselves cannot explain the popularity of design, nor the fixations and sometimes obsessions for some designs, which are what creates a niche for certain markets; nor is the human habit of copying, capable of giving success to a design that is not inherently nice, and stimulates people's interest, even if copying evidently disseminates it. And none of these aspects –as authentic as they are– seem satisfactory enough to compel the perpetuation of fine design after centuries and even millennia of its creation. The survival of forms is in the collective mind, the same mind that finds suggestions in decoration, or in the sensory quality of forms, or mirrors the shapes of its own imagination in clouds and in structures. Therefore, neither matching pieces nor the human inclination to copy seems the fundamental reason that motivates the design reapplication of culture. The answer is better situated in the culture's perception and identification of good and interesting design. The attraction of the copyist is not dissimilar from the attraction of buyers to enjoyable design. Enjoyable forms include structures and even instruments of natural aesthetic quality to the psyche.

2. Second Point. Everybody probably reacts to the visual impact of fine design, but overtly may quickly align their reaction to the 'expected' responses of a social milieu, which can either tone down or exalt the first impression after established norms. This results from previous conditioning that associates certain qualities in design with a given type of society. Thus, democratic values may cause us to reject the design of absolutism on the basis that it connotes the tastes and practices of past aristocracies, and we create ways to recognize this category by the presence of extravagance in design. This is a respectable social opinion but if it becomes a criterion to judge merit in artistic entities, then it is a very partial view. Designed entities have a merit of their own that is ultimately related to ingenuity and aesthetics. These socially-expected responses to design are also tied to what is trendy in society, as social values influence what is considered desirable or if preferable, stylish at a given point in time. Hence, at some point design may follow a social movement, for example the hippy or other trend, to soon die out together with an experiment in social living. But ranking design by socially trendy values is similar to acknowledging its merit only if it represents an expected outlook. This causes some of us to inhibit our freedom to like and enjoy good design and others to enjoy it but not express our enjoyment, as it may clash with what is trendy. Styles do fall out of favor, this is true, and we don't have to be talking about the styles of aristocracy to hear depreciatory words about any démodé style. However, while we may hear negative perceptions of historical design all the time, especially in reference to repetitive images and extravagance, we still happen to live amid the designs that never seem to die. The reality of what the public enjoys seems to be the reality we live by, and it doesn't easily die out.

Other types of circumstances related to the action of influences that are external to creativity on the way we rank design can be cited. It is important for example, to identify the types of circumstances that lead us to consider some work as artwork. One clear example of this type of circumstance might be an image that some persons feature prominently because it represents a worthy cause (or an activity that is pursued with passion, like a sport); however the form that represents the sports team (a sculpture of a mascot or a ball, for example) or something more grave and important, as the image used for the cure of an incurable disease, is usually not meant to be artistic. This example tries to illustrate in a simple way the very common effect of beliefs on design. There are many examples less evident than this one, where our emotional input for some architecture or artifact is also compelled by motivations external to the natural appeal of design. We can cite the action of friends sometimes on our choices and our love for our own regional design, both which can be influential in what we say that we like. The enjoyment of our region's crafts is partly tied to the visual domain of design, and it translates into ease and confidence which is felt in grasping and commanding its aesthetic syntax –something psychologically satisfying in relating to any design. Our region's forms, then –if they are of acceptable aesthetics– will commonly have emotional power over us. We may exemplify the design of our region as an emissary of its natural beauty and worth; however, it is still important to those who believe that the art of their own region is the best and the only one that counts, to discern that design merit is not a function of regional ties. The lasting images have been important to anybody and all mankind, regardless of their situation on the globe.

As with social values, major world religions judge that the human attachment to possessions is a delusion of the senses. We must first acknowledge that this is of course good wisdom, in the sense that material things don't give interior happiness in a human way, as material entities never take the place of friends. The wisdoms of religion may have indirectly introduced moderation in our views of design, and influenced the austere design periods of history. But religious advice has not rooted out our aesthetic capital, or just only temporarily so during an ascetic period of history; and religious centers collect treasures too. This tells that these commodities exert a driving power on the mind, despite the many warnings about their negative influence. Values may present ideological snags to our knowledge of design. Designed entities certainly may have high monetary value, but the point that is truly significant is that they directly or indirectly enhance life's experiences, thus, this leads to monetary value. There are good reasons for the human species to create and enjoy design and we can all provide tentative explanations: artistic creation improves the quality of life and it certainly enhances cultural life. Experiencing aesthetic buildings and spaces is exciting and it lifts up our mood. The upbeat effect of environments that impress people favorably is undeniable. All that design has to do to attain this benefit is to create places that enhance society's experiences.

3. Third Point. To an *avant garde* designer people's penchant for traditional styles can be thought to be unadventurous and conservative. Hence, the preference for traditional styles is often categorized today as a conservative choice. Even laypersons themselves have learned to categorize their own taste in a comparable way from ideas that circulate in the literature or art schools. Thus we can hear individuals who sometimes say about themselves "I am progressive in my ideas but conservative in my taste". These categorizations would not amount to much, if they didn't relegate sincere visual proclivity to the non-cognitive (or non-visual) world of values and by doing so closing a door to the visual behavior of people. Proclivity for old styles in untutored individuals is usually sincere. However what is agreeable to common persons can be inhibited in connoisseurs by predominant design values and the social behavior that is also inherent in the design professions. In true predisposition, there is a conscientious decision to choose a particular type of design after a belief or conviction, and this certainly can include both 'conservative' and 'progressive' choices. The point that this work makes is the simple fact that values can be peripheral to the sensory world or temporary, as what is conservative today used to be progressive in the past. Hence, what is *avant garde* will be classic one day too. Indeed some points of view of society that we take for genuine today are probably only temporal, or mere interpretations among other possibilities.

The evidence of the common proclivities in the public shows an abundance of historicist styles in residential housing, at least until today, and there seem to be cognitive grounds for this. Laypersons are easily captivated by craftsmanship, such as refined woodwork, moldings, roofing and some curves, not to mention the so-called 'nostalgic' looks of traditional architecture and even the eerie vernacular typologies of earlier times that also delight some people. In fact, it is interesting that a modernist architect such as Ludwig Mies is reported to have occupied a rather old fashioned apartment in Chicago at least for some part of his life. This does not mean at all, that people don't acquire contemporary items, as everybody brings into historicist homes technology and novel design finds, and there is a good amount of residential renovation involving high tech equipment that is done in contemporary lines. There is also ample evidence of a widespread intention in the general public to support contemporary arts and design. The public is interested in the subject, however, for some reason a large number of people have just not made the transition to living in housing that lacks traditional recollections, in particular those who reside in houses rather than urban apartments. In these latter ones we see more modernity. The causes of the success of old images in homes need to be objectively tackled and explained.

Part VI:

DISCUSSIONS AND CONCLUSIONS

Foreword to Part VI: Longings, Creativity and Fulfillment

In the triumph of structures there are other notions conveyed in addition to the many connotations discussed so far. The engineers who focus on building the bridges that link two pieces of land across an ocean may no longer remember that once, a wish made people study and explore a way to do it. Wishing that an aspiration could be real leads volition to research and invention may follow[196]. Discoveries and instruments are certainly the offspring of painstaking work, but also of wishes. Many artifacts and infrastructure that surrounds us once seemed unattainable, though they became quite common to our perception. Among the many substances of dreams were ideas such as flying or the idea of crossing distances in little time, or a road above a body of water, among many others. Before the age of airplanes and our viaducts and bridges, these structures were pure fantasy. Yet one day, together with outer space flight and sophisticated technology, they became a design obsession of modern times, until they were made real. No matter how plain or functional a structure, they have enabled some of the visions of wishes or great dreams. Sometimes the scientists behind an enabling technology have said this themselves[197].

Interesting and obsolete artifacts all tell a very vivid story. An old typewriter or pioneering sewing machine, an old telephone[198] or gramophone, can reflect the activity these artifacts performed, and some aspects of the society that created them. The long gone users of such products may be pictured in our minds when we contemplate these household items whose obsolescence makes them look curious to us. There is certainly something that is additional to the function and service that is liked which these entities once offered, which possibly plays a role in their glamour and the public's interest in them after they are obsolete; and whatever word we might choose, it should be something tied to the fulfillment of wishes, that the technological forms under view provided. From the Greek temple to the mystical Gothic, there were hopes, experiments and other mysterious relationships of the mind and creativity 'built' these into their appearances. This is the dialogue of the design of an age and the vision of how to build it, a permanent aspect of design. Possibly the ancient builders transferred the looks of rugs, wicker and timber construction into stone after sensing the hopes, discoveries and mysterious features built into them by earlier inventors.

It is clear that attraction to design that reaches people 30 to 500 generations after its image was created, may not be sensed exactly as contemporary aesthetics (in which case aesthetic implies beauty) but it can still be magnetic and elicit interest in the unrealized or subliminal domains and readings of the mind. The attraction of an antique for example, needs not reside in what we consider fine-looking today. Indeed, many antiques look perhaps bulky, possibly dark and maybe too decorative to our contemporary eyes. Yet when we shop for furniture, if we come across a fine piece of this type, it may appear interesting, and though we may believe that it is because it is a rarity, the piece awakens our subconscious by its appearance as peculiar (and sometimes even bizarre) as its shapes and forms in excess may look in our days. Certainly an antique that doesn't attract for its beauty, can be alluring for some other reasons: such as its extraordinary design (a hand-made craft or type that is no longer produced today) or the looks of its epoch, and because it elicits retrospection. The

---

[196] Longing must have compelled portable portrait painting of special persons and moods compel musical expression. From the wish to see and hear special persons and replay events, technology advanced to the creation of cameras, radios, gramophones and telephones.

[197] For example, the man behind the rocket to the moon, Von Braun, said he dreamed as a child to go to the moon.

[198] We have novel manual showers that evoke antique telephones.

wish of people to possess antiques is driven by the mind's exposure to fine examples, as a piece of music is appreciated by a certain amount of listening. A shopper who loves modernity may consider an antique unattractive, and hours later, the piece nevertheless comes back to mind, since it would surely make his interior space look special as a conversation piece. There is a transaction of form from designer to viewer and appearance tends to conquer the searching mind.

Chapter 21

The Sway of Possibility

Why should the way people and things look like matter at all? Probably the early twentieth century asked this question when the age advocated simplicity in design. One of the points was to make the world's psyche freer from its dependence on ornate looks, and this state of mind became the next road to be travelled. However, the proneness of design to become involved in establishing social castes remained the same in our new visual era, despite a new no-nonsense world of plain looks. The growth in population and knowledge brought even greater complexity to life and its transactions, and new forms of social stratification still took shape in our better world.

The appreciation of functional forms that are attractive in themselves –that is, they are aesthetic before treatment– additionally led designers to realize that embellishment is unnecessary in design. The very rough Roman aqueducts and cable bridges are fitting examples of interesting plain forms. Even rough stone gives a sense of strong texture to an arched row. Aqueducts offered a number of refinements in the form of suggestion of elements, however, which a cable bridge may usually not offer, since the twentieth century avoided these treatments because it has tried to be a different visual age. But the cable bridge is effortlessly aesthetic in its linear transparency and catenary lines, which are natural curves.

At Segovia, the aqueduct shows a suggestion of capitals (though squat and interpreted as necessary for scaffolding) at the spring of each arch, and a suggestion of pier bases and cornice underneath the water trough. Hence, it is a 100% utilitarian structure made with coarse stone that somehow has also been subject to the visual sensitivities of the mind who designed it. The opportunity to add these subtle enhancements by the Centurion who ordered it, and to taper the heights of the arched rows to make them look nice, was possible, and could not be missed by imagination[199]. This aqueduct is notwithstanding very different from a Baroque interior, but the entire appearance issue may, in the last analysis, be more a matter of degree rather than principle under the standpoint of the pre-twentieth century visual era.

---

[199] Fernandez Casado, 1972.

When architectural and religious philosophies have substituted decoration for undressed design, a reduction of embellishment has seemed both sensible and aesthetic in history too. This has happened before the twentieth century, though in a somewhat different way, such as in Herrera's *Desornamentado*[200] or, when religious orders or reform banned images and luxury as in Cistercian architecture. These works demonstrate the aesthetics of fairly plain structures –but touches of adornment were not lacking in the buildings of these movements. For example, there is a certain amount of decoration in the capitals and cornices at Cistercian abbeys and there is an exquisite smoothness in the stone used. This, combined with natural illumination that enhances the reflection of white surfaces, gives off a mystical light. But in spite of this subtle intended embellishment, the viewer can see the form of the pure mass so to speak and the beauty of the bare materials in plain structures, whose natural aesthetics were masterfully enhanced. Similar harmonies can be noticed in *El Escorial*. These and many other purer forms can be very captivating to viewers.

Upon realizing that pure forms can offer beauty (in themselves), an environment occupied by objects of only tectonic and utilitarian make-up can come to mind. Could the present and the future still want to move in the direction of an ornament-free world, as materials and technology allow the creation of such sophisticated forms that all the compensatory role of earlier ornament could be finally discarded? Judging from the millennia of the practice of ornament and the success of ancient forms (millennia after their conception) the educated guess should be that it will be very difficult to change the public mind, however, there surely might be an ornament-free modernized approach that could offer visions and fantasies that appeal to the public[201]. Nothing about the future can be established ahead of time in design. But anyone who harbors a futuristic vision may feel a strong wish to see what such an environment might look like. The very different and novel quality of such a place is likely to fire imagination. And the mind then begins to visualize a sleek and polished world and already begins to refine its contents. Therefore, the constructed world could be made of glass and metal, for example, as the early twentieth century Mies imagined. This metal will be beautiful, so it should better be copper-like, as many mid century high rises were, or burnished or brushed, as the later aesthetic steels, and exquisite rather than raw, as our age's preferences. Then, joints may be exposed or seamless, but experience in the end seems to have shown that seamlessness is the better option[202], and glass should be crystalline but some might be also tinted in brownish, orange and greenish, or be reflective….. These thoughts are all about enhanced appearance or embellishment; there should be no doubt about it. They are all visual refinements, and not unlike ornament itself. It is true that they have a basis on the aesthetics of functional forms, rather than on the aesthetics of flowers, cherubs, or mythical imagery. But any idea that they are not design enhancements leads us to the paradoxical understanding of these relationships that plague design education and critiques.

An ideal of embellishment-free design is on the other hand, not just a contradiction of the way the mind under cultural forces has worked until today, but also a simplified vision of creativity, not to mention that it is ultimately inhibitory or denying of an important instrument to creative individuals. Then, the desire to embellish is ingrained in the evolved social behavior of people.

---

[200] Influenced by Sebastiano Serlio's ideas.
[201] For example, some forms by Calatrava already achieve minimalism, structural soundness and biomorphism.
[202] We prefer seamless joints, and this brings to mind the fact that natural forms do not have seams.

Figure 200: Bar Faucet 2009 (Blanco).

The aesthetics of some materials is enhanced by polishing or coating the metal to a mirror shine. The handles are somewhat similar to that of a barrel of beer or a carpenter's vise, and though their recollection is that of a utilitarian mechanical device .. .("modern"), the spindle design is essentially historicist. Slenderness is enhanced relative to history's spindles.

Chapter 22

Summary: Interpretation and Illusion

Excessive multiplication of a design look is the enemy of its exclusivity. Design movements arise and attract, their looks are emulated and inundate the market; they fast reach a high point in the public interest and then they saturate people's attention and move out of the limelight. Proliferation of looks also once put out of style history's styles; however, the enduring images only lay dormant for some time, as if they carried an interesting nature or aesthetics that would elicit continued appreciation. In the scale of the millennia of human creative works people live for just a minute, comparatively speaking, and like butterflies or flowers some have taken pains to show good works and beauty, while others to make possible the dreams of society to be healthier and better. Some left a trace of what they thought aesthetic, and left us a vision that their minds were able to imagine and build, like the domes and the Gothic vaults. Later minds have silently understood those visions, grasped from their products, and generations have cultivated them. Their best images inspire connoisseurs and craftsmen to enjoy them and metaphorically 'replay' them as music is heard again. Music inevitably reflects strongly the replication phenomenon of the cultural or collective mind. It is a temporal art hence the similar dynamics to music that can be observed in the visual arts tells us about levels of universality in the arts, which in turn reflects perceptual modes of the mind. Some design seems to have been created in ways that made it flourish a second, third and fourth time, such as in the repetitions of classicism in the Renaissance, neoclassical age and the Victorian neo-Renaissance, or in the Gothic revival. Conscientious revivals were no doubt associated to a variety of factors or necessities, such as socioeconomic change, as in the Renaissance, or the archaeological discovery of Herculaneum and Pompeii that prompted neoclassicism, or the restoration of cathedrals that directed the field to the Gothic Revival; but all comes down to the rebirth of appreciation for significant creative work that having been dormant, but still felt, can take off from such other factors. The connoisseur, the amateur or the public that identifies some meaningful productions, and even though rarely discerns what is that meaning or substance in structures, know that it is present, and reproduce and preserve them. Both design reproduction and preservation reflect a passive but remarkable uphill effort against the dynamics of visual saturation that is inherent in styles, i.e., the demise of styles in time. This overall replaying of design against the odds is done by the collective mind we know as culture.

Creativity is one of the natural skills of the human species and it is used to provide a roof to families and society. Ancient structural forms had the important goal of equilibrating heavy elements by counteracting static forces, and make building forms work efficiently. The resulting early construction created structural forms. As structures and systems to build became sophisticated, creativity involved a search for fitting forms, as people have cognitive expectations for the conceptual and methodological instruments it utilizes, structures and mechanisms themselves being some of these implements. So culture sees and responds to forms, and the trades improve appearance in a most remarkable dynamics that may become directional. Architecture developed sophisticated social spaces. Meaning was attributed to structures by means of cultural feedback, as society grasped the equilibrated massiveness and sculptural forms of structures and used them as concepts associated to strength and the supportive function. This is most often an independent (and subsequent) development to the creation of structures. But everything seems potentially meaningful to culture and structures are one of those creations that seem to lead the list. Oral communication greatly helps its own richness by the use of analogies wherever images and forms can be resourceful to language, and design acquires connotations. The significance gained by buildings then suggests notions and this is shown in the uses of building images in connection to both religious and secular references that use the qualities of structures to refer to ideas of substance. Structures that turn out to be inherently aesthetic can become especially meaningful. By their aesthetics they invite the reapplication of their forms elsewhere, and by gaining meaning, they leave behind their original framework or system to become a free motif detached from its original context, -that is applied to the crafts. The creation of similar forms from an edifice's structure to its interiors, by collaborating trades is involved, but the structures that become significant are depicted outside their original context for apparently more critical purposes than this practical harmonizing of exterior and interior, and this involves the appreciation of their forms, as this is shown in the attachment of people to aesthetic entities and the reapplication of columns, domes, roofs and arches throughout history. Emulation of prototype structures through the crystallization of their essential elements (as in columns and pediment) indicates that there is interest and even esteem for such models.

There are boundaries and regulations to be followed in a style, and the mind respects such boundaries; but imagination will break through these constraints when it has a more advanced or interesting vision. There is also a changing 'faculty' in a prototype image, or an opportunity to modify it, in spite of its crystallized form. As buildings are emulated in new versions, they are designed somewhat differently, and may improve in some direction. The end result of buildings having a design that is derived from a prototype form is that they can be visually grasped as variations of a theme that the culture identifies collectively, cultivates and also transforms, miniaturizes and distorts, so that some forms evolve and flow in the process. This may lead to a new design movement. But culture also keeps some selected types of designs fixed at a certain valuable state, for their reputation or because adding or changing them disrupts the design perfection reached, and this effect compares with the type of vision that was put forth in the Renaissance.

Whereas some structures (as some mechanical instruments) are naturally agreeable and readily identified, many others are not. A plain arch is a satisfactory form as shown in its widespread application; this conclusion is from its seemingly perpetual visual resourcefulness even in cases where its wedges are not seen, as in the vernacular Mediterranean-styled whitewashed colonnades. In the shape of the arch and the hemisphere there are inherent forms of nature and culture, such as the vitalizing sun or the calming moon, hence, arcuated structures stimulate a recollection of these harmonious ideas. In fact arches can be reshaped slightly so as to also evoke the vital outline of the body. The recollections that exist in people's memory seem to play a role in an effortless aesthetic cognition that underscores the participation of the subconscious mind. Arches are enhanced aesthetically by repetition, and are presented as a rhythm.

Repetition of forms can be artistically expressive in culture, if achieved in amounts deemed suitable for a desired effect (rather than one that vulgarizes a form through excessive multiplication and coarse versions). Repetition is also observed in structural sequences and it is somehow present in experiencing a building from exterior to an interior of related forms and artifacts (as in the work of Robert Adam and Schinkel). Reiteration is common in many domains of culture, such as in cyclical commemorations, periodical rites and celebrations. We must also mention that reiteration is characteristic of renewal both in animated nature and in cyclical inanimate natural phenomena. Repetition is a theme of nature itself and our mind seems to have its own rendering of such natural dynamics except that it is not deterministic as natural cycles are, but it is done freely, for its resourcefulness and effect. In human oral and other communication, repetition can be observed as well. It is used because it is expressive. The visual repetition of traditions also seems to have a related flavor. Repetition can convey dedication and loyalty, for example, but also can be applied to express an emotion, as in music or poetry, or something urgent. These phenomena are motivations (or ways to do some things) that are inherent in the mind. The effect of spontaneous repetition, emulation and design migration belong in a different dimension than that of intention or even that of linguistic interpretation.

The arch's curvature gave a surrounding nature to domes and vaults. Enveloping design and frames, small and large, were abundant in ancient Rome. The medieval memory of these aesthetic forms seemed sometimes to distort the chiseled regular tectonic structural Roman arch into organic compartments for the human form. This compartment was present for example in the three-quarter circle arch (horse-shoe arch). The Middle Ages designed extensive replication of arches and vaults that produced expressive visual rhythms of structures. These included sculptural varieties and adaptations of vaults and arches that echo and celebrate the triumph of the structural vault, in small structural motifs. Artists found that arches and vaults were suitable as fitting frames for venerated sacred relics or portraits of holy persons. The aediculae show the use of a structural form eloquently, expressing the relationship of two parts, inanimate building and animated (living) figure, bringing out an unrealized and unnamed meaning of built form. They reflect a connotation of architecture that seems to be a constructed shell for the human form. The importance of the medieval vaults and the themes of matching forms such as aedicule and human figure, or scalloped arches and the brain's surface is this fundamental architectural or protective connotation. Could a fitting outline for a human figure, have been created by chance, or was a human form part of a process that gradually approximated it and was going to surface out of human imagination sooner or later? The fact that there are comparable representations of trefoil human outlines to the Western type, in India, the Middle East and Asia should be taken into account in trying to come up with a rationale, as this human outline could be an attempt to depict a natural concept of a protective cover that the mind can create under certain circumstances. In the West, this outline seems to have slowly developed from the structural arches. The earliest enabling arrangement may be that observed in the motif of a taller central arch flanked by two lower arches such as in a triumphal arch of Rome, and then this motif is present in the elevation of a basilica, and in a nave of a church. From here, to the sculptural trefoil of the Middle Ages, and centuries afterwards, to the hammer beam truss of the fourteenth and fifteenth century.

The creation of vaults had a profound effect in the communication of a visual notion of shelter that then was slowly formed until a complementary outline of the upper body in the trefoil shape and the hammer-beam-truss appeared. Some decorative vaulted forms, such as the Gothic aediculae, show effort in reinterpreting the church's nave in sculpture, that is, in art, possibly indicating a high appreciation from the part of craftsmen for the structural vault –whether the age perceived it as a great construction accomplishment or as a sacred space. Vaulted miniatures depicted in association to human figures had the effect of communicating this notion even more clearly than the full-scale structures by themselves. Sometimes lobed vaults became tiny such as the *mocarabes* of Islam which are like miniature niches in their form, and became profusely multiplied in a way that reflects a membrane-like natural formation suspended vertically from a ceiling (as in the Islamic Nasrid interior decoration of the fourteenth to fifteenth century). Other ceilings such as the vaulting of the Friday Mosque of Morocco, the lobed arches, and the late Gothic development that uses pendants hanging down from a light colored vault, such as some of the later vaults of England, reflect parallels with the outline of the brain, sometimes even to a minute level of detail.

To the field of psychology, surrounding or enveloping forms in unconscious dream imagery symbolize maternal functions[203]. Perhaps this maternal 'principle' of the mind is active in the allegories of architecture such as ecclesia and synagogue or city and nation allegories as the statue of Liberty (a classical styled allegory of a nation portrayed as a strong matron) that are depicted in metaphoric art as a female figure. The Mother archetype has been painstakingly discerned by the long experience of psychologists from innumerable patient visions of unconscious imagery. When we look at design, we have a materially-carved or molded, revealing product of the mind, that is available for us to discover, rather than a memory of a vague unconscious dream, and the design informs its conceptual inspiration by displaying its configuration. We have to understand in particular the outcomes that the mind does not intend, but it nevertheless designs, so as to plumb its dynamics free from a priori convictions. We ought to look too, at a possible biology in some of the most vital of the mind's creations. Not in vain the brain *is* a biological organ and the ability to understand is too, even if it surpasses the very ways of nature, and works independently from it; its make-up is not changed by this fact. For example, and this is just a conjecture at this point, the whole picture and behavior of structural motifs could speak of a primitive logic in creativity that may go back to the dramatic necessity to seek for, and provide, fitting forms of protection that can remedy and rectify human vulnerability, where architecture and mantle blend. Once a satisfactory method is created, culture repeats it and the collective subconscious does not want to leave it behind, and continues to wish to produce it, not clearly understanding why. The creative search must have first come as a strong interest (like an obsession to build), that existed in any pre-human mind that busied itself in the creation of its earliest instruments. This interest does not cease by arriving at a fitting design, but this interest keeps searching for interesting forms. The subconscious mind had the ability to arrive at a very fitting form, as the design examples of a matching interior concavity for the upper human figure reveal, but on the other hand the creative interest never ceases.

The forms of objects and vestments that are ceremonially associated to a building are influenced by the edifice's character and harmonious phenomena. The fact that architectural motifs are passed to head gear seems to refer to the protective sensory role of buildings, in particular, to the head. Conversely, the embellishment of architecture can sometimes convey something of the human form. The two conceptual expressions are likely connected in the mind's subconscious and sensory cognition.

---

[203] C. G. Jung's *Aion*.

Understanding the use and manipulation of materials to please the psyche's searches is to understand the sensory qualities of design. The sensory qualities of materials have been essential helpers of form and suggestion in design. These qualities are part of embellishment or treatment, and may be particularly important when a design approaches bio-mimesis. Many important building examples indicate that interiors have been deemed optimal when smooth inside. This derivation may find an example in the shell's interior, but also our brain's interior is smooth. Then, marble, plaster, porcelain, mother of pearl, glass and equivalent glossy and dense materials are suitable for an inner search of our psyche for soft lining materials. Crafted wood has also had a long development in the sensory polished direction that is evident in curvy and organic furniture.

Society holds on to favorite images steadfastly. It is as if we find ourselves involved in a collective theme of keeping the images of our cognitive interests. Should ancient architecture wear off with the passing of time, culture may continue to reproduce their images in the crafts. As we mentioned in the introduction, this activity has been happening since ancient people reproduced archaic details in their newer architecture. The best images are the best civilizing remnants of an age, and culture holds on to them. This topic probably has alternate important and elaborate interpretations, but in the structures that are built with complementary outline to that of the human upper body, we find a key cognitive metaphor of what architecture can be to the mind, and this could lead us to one answer, even though there may be others. This metaphor envisions architecture almost as a natural case that is designed for the human form, which can also take the form of the natural cover of the mind. The mind arrives at its own form in some design. The cognitive exposure to the shell may have had something to do with this solution too. The scallop shell has been a creature of intense interest to all early human cultures, especially as a symbol of protection. It has also been a revealing form for the 'structural' conception of niches and gored domes. The architectural form (or the structural motif) of a funerary object, including the shell niche, reveals an association between architecture and that solemn pathos of making last what is left of a revered person. So does the motif of a gored dome in a reliquary –a ciborium by another name. Indeed we might accurately acknowledge that buildings have made our species last until today.

The imagery showing together people and vault, or person and niche, and, the costume designs using architectural configurations, are motifs about the relationship of architecture (a constructed shelter) to a person. In all cases, people use a meaningful structural form to communicate some connotations that architecture and structures can suggest.

The dynamics of the design of human appearance offers some points of comparison to the dynamics of architecture and interiors. In the use of structural recollection in costume, as in the use of the anatomy in a vault or dome –as that of the Pantheon– the historical user was most likely not aware of most of these structural metaphors that were built or that he or she wore[204]; however, visually, s/he was probably accustomed to several of the correspondences of these different forms. Quite certainly the entire phenomenon of resorting to structural forms –and conversely, of modification of structures by adornment, so as to make structures recall some concepts– has unintentional components that sometimes are not easily identified, especially by our verbal-driven side of the mind. This is undeniable in several examples. Still there are some people who hesitate to accept that such parallels are arrived at (and perpetually kept) by

---

[204] Although some parallel ritual forms of India, the Far East and South East Asia (in pagodas, temples, dance costumes, theater) are obviously designed today in full awareness of parallelism.

(45) Rich, Anthony, *A Dictionary of Roman and Greek Antiquities* with Nearly 2000 Engravings on Wood From Ancient Originals Illustrative of the Industrial Arts and Social Life of the Greeks and Romans, Longmans, Green and CO., London, 1884; revised 1893.

(46) Sagredo, Diego de, *Medidas del Romano*, Toledo 1526. Reproducción del original, Patronato del Instituto del Libro Español, Ejemplar N. 114, con privilegio, Madrid, 1946.

(47) Schiller, Gertrud, *Iconography of Christian Art*, Volumes I and II, New York Graphic Society Ltd., 1972.

(48) Snodin, Michael, *Karl Friedrich Schinkel: A Universal Man*, Yale University Press and Victoria and Albert Museum, New Haven and London, 1991.

(49) Schwenke, F., *Gründerzeit. Möbel und Zimmer-einrichtungen*. Verlag von Ernts Wasmuth, Berlin, 1881. Republished in 1985.

(50) *The National Gallery Supplement of Art News*, 1946, The Art Foundation Inc., New York.

(51) Rolf, *Klassizismus und Romantik, Architektur, Skulptur, Malerei, Zeichnung*, 1750–1848. Könemann, Koln, 2000.

(52) Von Egle, J., et al., *Baustil und Bauformenlehre*, I. T. H. Schäffer, Hannover, 1905. Republished in 1995.

(53) Von Engelhorn, Verlag, *Musterbuch für Kunstschlosser*, Stuttgart, 1885. Reprint-Verlag-Leipzig; Verlagsarchives. Reprint der Originalausgabe. Lektorat: Andreas Bäslach, Leipzig.

(54) Von Wölfel, Wilhelm, *Brunnen-Brücke-Aquädukte*, Berichte zum Bauen in der Antike, Verlag fur Architektur und technische Wissenschaften, GmbH, Berlin, Ernst & Sohn 1997.

(55) Wirtz, Rolf C. *Florenz*, Kunst und Architektur, Könemann, Köln, 1999.

(56) Whiton, Sherill, *Interior Design and Decoration*, J. B. Lippincott Company, NY, 1974.

(57) Zerbst, Reiner, *Antoni Gaudi, Benedikt Taschen*, GmbH, Köln, 1993.

(58) Zimmer, Heinrich, *Myths and Symbols of Indian Art and Civilization*. Bollingen Series VI, Princeton University, 1992.

Previously Consulted for this Topic

(59) Bramfels, Wolfgang, *Die Welt der Karolinges and ihre Kunst*, Verlag, München, 1968.

(60) Branner, Robert, *Chartres Cathedral*, Thames and Hudson, London, 1969.

(61) Bruhn, Wolfgang, and Max Tilke, *A Pictorial History of Costume*, Hastings House Publishers, New York, 1973.

(62) Burgess, S. C. and King, A. M., *The application of animal forms in automotive styling*, The Design Journal, 2005.

(63) Busch, Harold and Bernd Lohse, *Vorromanische Kunst and ihre Wurzeln*, Umschau Verlag, Frankfurt am Main, 1965.

(64) Campbel, Donald T., "Evolutionary Epistemology", in *The Philosophy of Karl Popper*, Schilpp ed., 1970.

(65) Cherry, Colin, *On Human Communication*, The MIT Press, Cambridge, Mass., 1975.

(66) Cowan, Henry J., *A History of Masonry Concrete Domes in Building Construction*, Structures Report SR11., Department of Architectural Science, University of Sydney NSW, Australia, 1977.

(67) Carpenter, Rhys, "The Aesthetics of Greek Architecture", in *Readings in Art History*, V.I., Screibners, New York, 1976.

(68) Duby, Georges, *L'Europe au Moyen Age: Art Roman, Art Gotique*, Arts et Metiers graphiques, Novembre, 1979.

(69) Dixon, Arthur, "Some Observations on Ancient and Modern Buildings", Proceedings of Allied Societies, The Birmingham Association, *Journal of the Royal Society of British Architects*, Mar. 5, pp. 297–300, 1896.

(70) Dondis, Dondis A., *A Primer on Visual Literacy*, The MIT Press, Cambridge, Mass, 1973.

(71) Dorn, Harold, and Robert Mark, "The Architecture of Christopher Wren", *Scientific American*, vol. 245, no. 1, July 1981.

(72) Emerson, William, "The Desirability of Official Control Over Architecture in Our Towns and Cities", Tuesdays Proceedings, *Journal of the Royal Society of British Architects*, Jun. 30, pp. 408–409, 1900.

(73) Focillon, Henri, "The Eleventh Century and the Romanesque Church", *Readings in Art History*, V. I., Screibners, New York, 1976.

(74) Foster, Mary L., and Stanley Brandes, editors, *Symbol as Sense: New Approaches to the Analysis of Meaning*, Language Thought and Culture Series, International Symposium on Symbolism, Austria, July 1977, Academic Press, Berkley, 1980.

(75) Frankl, Paul, "The General Problems of the Gothic Style", *Readings in Art History*, V.I., Screibners, New York, 1976.

(76) Frankl, Paul, *Handbuch der Kunstwissenschaft: Mittelalters. Die Frühmittelhalterliche und Romanische Baukunst*, Akademische Verlagsgesellschaft, Athenaion, M.B.H., Wildpark, Postdam, 1926.

(77) Gorsline, Douglas, *What People Wore*, Crown Publishers, New York, 1952.

(78) Hamann, Richard, *Deutsche und Französische Kunst im Mittelalter*, Kunstgeschichtliches Seminar, A. Lahn, Marburg, 1922.

(79) Hamlin, A. D. F., *A History of Ornament*, Ancient and Medieval, The Century Co., New York, 1916.

(80) Hart, Franz, *Kunst und Technik der Wölbung*, Verlag Georg D.W. Callwey, München, 1965.

(81) Hell, Vera and Hellmut, *Die Grosse Wallfahrt des Mittelalters Kunst an den Romanischen Pilgerstrassen durch Frankreich und Spanien nach Santiago de Compostela*, Verlag Ernst Wasmuth, Tübingen, 1964.

(82) Holmdahl, Gustav, et al., editors, *Gunnar Asplund Architect: 1885–1940*, Plan Sketches and Photographs, Svenska Arkitekters Riksförbund Publisher. Copywright Abtidskriften Byggmästaren, Stockholm, 1950.

(83) Hurlock, Elizabeth B., *The Psychology of Dress: An Analysis of Fashion and its Motive*, Benjamin Blom Inc., Publishers, New York, 1971.

(84) Jairazbhoy, R. A., *An Outline of Islamic Architecture*, Oxford University Press, 2003.

(85) Joacobi, Jolanda, *Complex/Archetype/Symbol in the Psychology of C.G. Jung*, Translated by R. Manheim, Pantheon Books, New York, 1959.

(86) Kaschnitz-Weinberg, Guido, Freiherr von, *Die Mittelmeerischen Grundlagen der Antiken Kunst*, August Osterrieth, Krankfurt A.M., 1944.

(87) Kaschnitz-Weinberg, Guido, Freiherr von, *Das Schöpferische in der römischen Kunst. Römische Kunst I*, rowohlts deutscher enzyklopädie, Rowohlt Taschenbuch Verlag GmbH, Reinbek bei Hamburg, September 1961.

(88) Kaschnitz-Weinberg, Guido, Freiherr von, *Zwischen Republik and Kaiserzeit. Römische Kunst II*, Rowohlt, Reinbek bei Hamburg, November 1961.

(89) Kaschnitz-Weinberg, Guido, Freiherr von, *Die Grundlagen der republikanischen Baukunst. Römische Kunst III*, Rowohlt, Reinbek bei Hamburg, May, 1962.

(90) Kaschnitz-Weinberg, Guido, Freiherr von, *Die Baukunst im Kaiserreich. Römische Kunst IV*, Rowohlt, Reinbek bei Hamburg, January 1963.

(91) Kaschnitz-Weinberg, Guido, Freiherr von, *Kleine Schriften zur Struktur*, Ausgewahlte Schriften Band I, Verlad Gebr. Mann, Berlin, 1965.

(92) Kaschnitz-Weinberg, Guido, Freiherr von, *Mittelmeerische Kunst Eine Darstellung ihrer Strukturen*, Band III, Deutsches Archaologisches Institut, Verlag Gebr. Mann, Berling, 1965.

(93) Katzenellenbogen, Adolf, *The Sculptural Programs of Chartres Cathedral: Christ. Mary. Ecclesia.*, The Johns Hopkins Press, Maryland, 1959.

(94) Kessler, Christel, *The Carved Masonry Domes of Medieval Cairo*, Art and Archaeology Research Papers, The American University Press in Cairo, London, 1976.

(95) Krautheimer, Richard, *Early Christian and Byzantine Architecture*, Pelican History of Art, Penguin Books, Maryland, 1965.

(96) Krautheimer, Richard, "Introduction to an 'Iconography of Architects' in the Middle Ages", *Jour. of the Warburg and Courtauld Institutes*, vol. 5, pp. 1–33 ,London, 1942.

(97) Langer, Susanne K., *Mind: An Essay on Human Feeling*, The Johns Hopkins Press, Maryland, vol. 2, 1972.

(98) Laver, James, *Style in Costume*, Keeper of Prints and Drawings in the Victoria and Albert Museum, Oxford University Press, London, 1949.

(99) Lehmann, Edgar, *Der Fruhe Deutsche Kirchen Bau*. Die Entwicklung seiner Raumanordnung bis 1080. Tafel I and Tafel II. Deutscher Verein fur Kunstwissenschaft Band 27, Berlin, 1949.

(100) Lesser, George, *Gothic Cathedrals and Sacred Geometry*, Vol. I, Alec Tiranti, London, 1951.

(101) Lewis Kausel, C., "The fundamental expression of an architectural motif", *SM Thesis*, Department of Architecture, Massachusetts Institute of Technology, 1982.

(102) Lewis Kausel, C., "The image of architecture in objects", *Journal of the Interfaith Forum of Religion, Art and Architecture* (IFRAA), vol. 1, Fall, 1986.

(103) Lewis Kausel C., and A. P. Julian, editors, *Santiago Calatrava, Conversations with Students*; the MIT lectures, Princeton Architectural Press, 2002.

(104) Mallgrave, H.F., *Gottfried Semper, Architect of the Nineteenth Century*, Yale University Press, 1996.

(105) MacDonald, William L., "Roman Architecture", *Readings in Art History*, Vol. I, Screibners, New York, 1976.

(106) Mark, Robert, "Tension in the Cathedral", Gothic Beauty, was it Form or Function, by M. Cohalan, *Science*, vol. 81, pp. 32–41. December 1981.

(107) Nelson, Betsy, *Structural Principles Behind the Failure of Gothic Cathedrals*: The Problems of the Central Crossing. Gift of Professor Traum to MIT Libraries, Spring 1971.

(108) Nuere, E., *La Carpintería de Armar Española*, Instituto de Conservación y Restauración de Bienes Culturales (ICRBC), Madrid, Spain, 1989.

(109) Oakeshott, Walter, *Classical Inspiration in Medieval Art*: Rhind Lectures for 1956, Chapman and Hall, London, 1959.

(110) Oursel, Raymond, *Living Architecture: Romanesque*, Grosset and Dunlap Inc., Fribourg, Printed in Switzerland, New York, 1967.

(111) Panofsky, Erwin, *Studies in Iconology*, Oxford University Press, New York, 1939.

(112) Panofsky, Erwin, *Gothic Architecture and Scholasticism*, Archabbey Press, Latrobe, Pennsylvania, 1951.

(113) Panofsky, Erwin, *Tomb Sculpture: Its Aspects from Ancient Egypt to Bernini*, Abrams, New York, 1964.

(114) Philipson, Morris, *Outline of Jungian Aesthetics*, North American University Press, 1963.

(115) Popper, Karl, "The Rationality of Scientific Revolutions", in *Problems of Scientific Revolution*, R. Harre, ed., Clarendon Press, Oxford, 1974.

(116) Portoghesi, P., *Natura e Archittetura*, Skira editore, Milano, 1999.

(117) Rasmussen, Steen Eiler, *Experiencing Architecture*, The MIT Press, Cambridge, Mass., 1972.

(118) Rowe, Collin, and F. Koetter, *Collage City*, The MIT Press, Cambridge, Mass., 1978.

(119) Solokorpi, Asko, *Modern Architecture in Finland*, Praegen Publishers, New York, 1970.

(120) Schwartz, Robert, "Representation and Resemblance", *The Philosophical Forum*, vol. v, no. 4, 1975.

(121) Schwarzenski, Hanns, *Monuments of Romanesque Art*. The Art of Church Treasurers in North-western Europe, Faber and Faber, London, 1967.

(122) Smith, E. Baldwin, *The Dome*, Princeton University Press, New Jersey, 1950.

(123) Storr, Anthony, *Modern Masters*, C.G. Jung, ed., Frank Kermode, The Viking Press, New York, pp. 33–35, 1973.

(124) Summerson, John, *The Classical Language of Architecture*, The MIT Press, Cambridge, Mass.

(125) Summerson, John, *Heavenly Mansions and Other Essays on Architecture*, The Norton Library, W.W. Norton and Company Inc., New York, 1963.

(126) Thompson, D'Arcy Wentworth, *On Growth and Form*, an abridged edition edited by John Tyle Bonner, Cambridge at the University Press, 1961.

(127) Torres Balbás, "Arcos Lobulados y Nichos", *Al-Andalús*. 1956.

(128) Venturi, Robert, et al., *Learning From Las Vegas*, The MIT Press, Cambridge, Mass. 1972.

(129) Volbach, Wolfgang, and M. Hirmer, *Early Christian Art*, Abrams Inc., New York, 1962.

(130) Warth, O. *Die Konstruktionen in Stein*, J.M. Gephard's Verlag, Leipzig, reprinted in 1993 by Verlag Th. Schäffer, Hannover, 1903

(131) Warth, O., *Die Konstruktionen in Holz*, J.M. Gephard's Verlag, Leipzig, reprinted in 1995 by Verlag Th. Schäffer, Hannover, 1900.

(132) Wilde, James, "In New York: Mortar and the Cathedral", American Scene, *Time* magazine, May 1981.

(133) Wittkower, R., *Architectural Principles in the Age of Humanism*, with a new introduction by the author, The Norton Library, 1971.

(134) Wrede, Stuart, *The Architecture of Erik Gunnar Asplund*, The MIT Press, Cambridge, Mass., 1980.

(135) Yarza Luaces, J., *Arte y Arquitectura en España*, pp. 500–1250, 1979.

Guidebooks:

(136) Baer, Winfried, and Ilse, et al., *Charlottenburg Palace*, Berlin, Musees et Monuments de France, Fondation Paribas, 1995.

(137) Hall, M., Cambridge, The Pevensey Press, David and Charles plc., Devon, 1995.

(138) Bombe, Walter, *Perugia*, Verlag F.A. Seeman, Berumte Kunststatten Band 64, Leipzig, 1914.

(139) Cardi, Enzo, *Montalcino museo civico, "Musei d'Italia-Meravigli d'Italia,"* Edizioni Calderini, Bologna, 1972.

(140) Fling, Karl, (co-editor), Alvar Aalto B and 1922–1962, *les Editions d'Auchitecture* Artemis, Zurich, 1970.

(141) Garzelli, Annarosa, Orvieto museo dell opera del Duomo, Musei d'Italia-Meraviglio d'Italia, Edizioni Calderini, Bologna, 1972.

(142) Huyghe, René editor, *Larousse Encyclopedia of Byzantine and Medieval Art*, Prometheus Press, New York, 1958.

(143) Peroni, Adriano, *Pavia museo civici del castello visconteo, Musei d'Italia-Meraviglie d'Italia*, Edizioni Calderini, 1975.

(144) Zovato, Paolo Lino, *Portogruaro, museo nazionale concordiese, Concordia, scabi, battisterio, Summaga, abbazia, Sesto al Reghena, abbazia, Caorle, Musei d'Italia-Meraviglie d'Italia*, Edizioni Calderini, Bologna, 1971.

Topic Index

Index of Themes

Colour in Art, Design and Nature

Edited by: C.A. BREBBIA, Wessex Institute of Technology, UK; C. GREATED, University of Edinburgh, UK and M.W. COLLINS, (Formerly Brunel University, UK)

The full-colour works in *Colour in Art, Design & Nature* form a striking contribution to the commonwealth of colour studies and to viewing science and art as complements rather than opponents (a la C. P. Snow's Two Cultures). This book is ambitiously inter-disciplinary and will appeal to academics and practitioners from a variety of fields. Colour and inter-disciplinarity go hand in hand. This so often involves the authors leaving the comfort zone of their original speciality and striving for excellence in another. It seems that our perceptions of aesthetics and beauty must be very flexible indeed so as to find absolute opposites equally fascinating.

The book involves four main types of contributions, defined in terms of the authors themselves. First, there are contributions by biologists. Second, the largest section is by practising artists. Third, there are two engineering-based contributions. Finally, there are contributions that address some of the historical proponents of colour theory and art.

CONTENTS: Animal camouflage: biology meets psychology, computer science and art; Lusciousness, the crafted image in a digital environment; The diversity of flower colour: how and why?; Sensations from nature; Goethe, Eastlake and Turner: from colour theory to art; Zvuk; Time and change: colour, taste and conservation; Thermo-hydraulics, colour and art; Nature's fluctuating colour captured on canvas?; On the use of colour in experimental fluid mechanics; Maxwell's first coloured light sources: artists' pigments; Past present and future craft practices project; Figuring light: colour and the intangible; Gaelux™; Colour in the countryside buildings, landscapes, culture; Developing the CREATE Network in Europe; Colour, light and sacred spaces; Analysis of the use of yellow in seventeenth-century church interiors.

ISBN: 978-1-84564-568-7   eISBN: 978-1-84564-569-4
Published 2011 / 154pp / £59.00/US$118.00/!83.00

WIT eLibrary

Home of the Transactions of the Wessex Institute, the WIT electronic-library provides the international scientific community with immediate and permanent access to individual papers presented at WIT conferences. Visitors to the WIT eLibrary can freely browse and search abstracts of all papers in the collection before progressing to download their full text.

Visit the WIT eLibrary at
http://library.witpress.com

WITPRESS *...for scientists by scientists*

Design and Nature VI

Comparing Design in Nature with Science and Engineering

Edited by: **C.A. BREBBIA**, *Wessex Institute of Technology, UK and* **S. HERNÁNDEZ**, *University of A Coruña, Spain*

Throughout history, many leading thinkers have been inspired by the parallels between nature and human design, in mathematics, engineering and other areas. Today, the huge increase in biological knowledge and developments in design engineering systems, together with the growth in computer power and developments in simulation modelling, have all made possible more comprehensive studies of nature. These developments have been reviewed biennially in a series of conferences first held in 2002. This book contains the papers presented at the latest conference in the series.

Topics include the following: Mechanics in Nature; Nature and Architecture; Natural Materials and Processes; Solutions from Nature; Biomimetics and Bio-inspiration; Biocapacity; Education in Design and Nature; Competition in Nature; Biological Engineering; Constructional Theory; Locomotion in Nature; Gravitational Biology; Self-sustaining Environments.

The book will be of interest to researchers from around the world who work on studies involving nature and its significance for modern scientific thought and design.

WIT Transactions on Ecology and the Environment, Vol 160
ISBN: 978-1-84564-592-2   eISBN: 978-1-84564-593-9
 2012 / apx 600pp / apx £258.00/US$516.00/€361.00

Compliant Structures in Nature and Engineering

Edited by: **C.H.M. JENKINS**, *Montana State University, USA*

Nature is the grand designer and human engineers have taken great motivation from it since the earliest of times. This book celebrates structural compliance in nature and human technology. Examples of compliant structures in nature abound, from the walls of the smallest cell, to the wings of the condor, to the tail of the gray whale. The subject of compliant structures in nature and engineering is timely and important, albeit quite broad and challenging. A concise summary of the important features of these interesting structures, this volume demonstrates, wherever possible, a mapping between naturally compliant structures and the promise and opportunity commensurate in human engineering.

Series: Design and Nature, Vol 5
ISBN: 978-1-85312-941-4
Published 2005 / 296pp / £107.00/US$214.00/€150.00

WITPRESS ...*for scientists by scientists*

Eco-Architecture IV

Harmonisation between Architecture and Nature

*Edited by: **C.A. BREBBIA**, Wessex Institute of Technology, UK*

Containing the proceedings of the latest in a series of conferences on the emerging topic of eco-architecture, this book presents the newest research in the field.

Eco-architecture requires that buildings be in harmony with nature, including their immediate environs. Locations, siting and orientation, as well as the materials used, should be chosen based on ecological appropriateness. Practitioners make every effort to minimise the use of energy at each stage of a building's life cycle, including that embodied in the extraction and/or fabrication as well as the transportation of the materials used and their assembly into the building. There is even consideration given to the ease and value of changing use of a building and component recycling when the building's life is over. Designers may also carefully control the energy required for building maintenance, not to mention lighting, heating and cooling, especially when the energy consumed is related to greenhouse gas emissions. Passive energy systems such as natural ventilation, summer shading and winter solar heat gain also play a role, as do alternative sources of energy for heat and electricity, e.g. solar and wind power.

Papers presented cover topics such as: Ecological and Cultural Sensitivity; Design by Passive Systems; Life Cycle Assessment; Quantifying Sustainability in Architecture; Resource and Rehabilitation; Building Technologies; Ecological Impact of Materials; Durability of Materials; Adapted Reuse; Carbon Neutral Design; Education and Training; Case Studies; New Architecture Frontiers; Art and Craft; Quality in Architecture; Temporary Architecture; Selection.

WIT Transactions on Ecology and the Environment, Vol 165

ISBN: 978-1-84564-614-1   eISBN: 978-1-84564-615-8
2012 / apx 600pp / apx £258.00/US$516.00/€361.00

WITPress
Ashurst Lodge, Ashurst, Southampton,
SO40 7AA, UK
Tel: 44 (0) 238 029 3223
Fax: 44 (0) 238 029 2853
E-Mail: witpress@witpress.com

All prices correct at time of going to press but subject to change.
WIT Press books are available through your bookseller or direct from the publisher.

Find us at
http://www.witpress.com